Chemical Engineering in Oil and Gas Production and Other Complex Subsurface Processes

AA003218

Topical Conference at the 2010 AIChE Annual Meeting

Salt Lake City, Utah, USA
7-12 November 2010

ISBN: 978-1-61782-142-4

Chemical Engineering in Oil and Gas Production and Other Complex Subsurface Processes

Topical Conference at the 2010 AIChE Annual Meeting

Salt Lake City, Utah, USA
7-12 November 2010

ISBN: 978-1-61782-142-4

AIChE

Printed from e-media with permission by:

Curran Associates, Inc.
57 Morehouse Lane
Red Hook, NY 12571

Some format issues inherent in the e-media version may also appear in this print version.

Printed by Curran Associates, Inc. (2011)

AIChE
3 Park Avenue
New York, NY 10016-5991

Phone: (203) 702-7660
Fax: (203) 775-5177

www.aiche.org

Additional copies of this publication are available from:

Curran Associates, Inc.
57 Morehouse Lane
Red Hook, NY 12571 USA
Phone: 845-758-0400
Fax: 845-758-2634
Email: curran@proceedings.com
Web: www.proceedings.com

TABLE OF CONTENTS

(24a) Data Requirements for Heavy Oil Production - A Reservoir Fluid Perspective

Shawn D. Taylor and John Ratulowski, DBR Technology Center, Schlumberger, Edmonton, AB, Canada

There is a spectrum of methods available for the production and transportation of heavy oils from a variety of different environments (e.g. arctic, onshore, offshore, urban). The production methods may be grouped into one of three main categories; primary methods such as CHOPS, thermal methods including combustion and steam injection, and non-thermal methods such as VAPEX. These methods may be used on their own in or in combination, such as solvent assisted steam injection processes. In each of these cases, some level of understanding of the reservoir fluids is required to select and optimize the production process. As a general rule of thumb, techniques requiring alteration of the fluid properties during production require a greater understanding of the fluid properties.

From an industry perspective, there are a few key challenges that are common to understanding the heavy oil phase behavior and fluid properties for all production processes. The first challenge is deciding what samples to collect. The choice of sampling is determined to some extent on the type of processes (and the related data requirements) under consideration for a particular reservoir. In addition, there is a growing awareness that heavy oil fields often involve fluids with a range of fluid properties, where the variation in properties may be both vertical and aerial. Therefore, there is a need to develop new sample selection approaches that provide the necessary samples that meet the data generation requirements of the production processes of interest as well as to allow identification and quantification of potential fluid properties variations within a heavy oil field.

After selection of sampling locations, there is the challenge of actually acquiring the fluid samples. Ideally, samples would be acquired directly from the reservoir using fluid sampling tools. Unfortunately, due to the low mobility of heavy oil fluids in a reservoir, such tools are often ineffective and most fluid samples must be acquired at the surface when the wells are already flowing or through extraction of a small volume of fluid from a limited number of cores. As a result, thermodynamic models for reservoir simulation or transport will be developed using estimated fluid properties since key data may be not be measurable with the available samples, the data may not provided in time, or there is simply a lack of representative samples.

Once in the laboratory, there are challenges related to the handling of the samples and subsequent measurements. It most case the fluid samples contain some level of contamination, which may include drilling mud, fine solids, and emulsified water, that can alter the measured fluid properties. While there are options available to remove such contaminates, there is no reliable standard approach for 'cleaning' that ensures the samples remain representative. Thus, there is a need to develop a standard method or suite of standard methods for preparing 'representative' fluid samples for measurement. The final challenge is the ability to routinely and consistently characterize heavy oil fluids. While some data can be readily generated using existing standard methods, either directly or through minor modifications to conventional oil procedures, there are a number of key data needs where there is no 'standard' methodology. A simple example is the whole oil composition for heavy oils, where the options are often limited to methods that do not offer enough detail about the composition (e.g. SARA) or cannot resolve a significant portion of the composition (e.g. GC-based methods). A more detailed knowledge of the heavier components is needed for fluid modeling. A solution to this example may be a suite of standardized compositional approaches that address the requirements of each type of production process.

This presentation will provide a brief overview of the types of data required by different production processes and attempt to identify areas of potential technology development in the selection, acquisition, handling, and analysis of heavy oils.

See more of this Session: Heavy Oil and Other Challenging Resources I
See more of this Group/Topical: Topical D: Chemical Engineering in Oil and Gas Production and Other Complex Subsurface Processes

(24b) New Directions for the Oilsands and Heavy Oil Industry Enabled by Fundamental Research

John M. Shaw, Chemical and Materials Engineering, University of Alberta, Edmonton, AB, Canada

Hydrocarbon production and refining have entered their second century as global, industrial scale processes. There is a tendency to treat them as sunset industries with only limited research and development needs and potential, as large thermophysical property databases have been created, and a panoply of experimental techniques, and commercial computational tools have been available for some time. While it can be argued that for natural gas and light oil processing, this is so, even for these cases critical exceptions arise particularly with respect to phase behavior and physical property prediction quality that affect process safety and optimization [1]. For unconventional hydrocarbon resources (e.g.: heavy oil and bitumen), a growing segment of the hydrocarbon energy and petrochemical industries, this is certainly not the case. Customary experimental and computational tools are of limited value because bitumen and heavy oil include nano-dispersed and self-assembled phases, are opaque to visible light, include solid phases, and comprise large molecules that are ill-defined. Process identification, design and optimization present formidable challenges for engineers and researchers. Phase equilibria, and transport properties [2, 3] are not readily measured or predicted. Chemical reaction schemes are not readily optimized. Apparently simple measurements such as viscosity, a key transport property in production and pipelining, have become an uncertain and debated topic [3]. The number and nature of phases present and the enthalpy of mixing associated with bitumen + n-alkanes, a mixture class which arises in paraffinic deasphalting and blending, remain unresolved. Much work is conducted blind or by analogy and it is not clear whether existing processes, from production to separation and refining are appropriate or have been optimized with respect to capital or operating cost, greenhouse gas emission, efficiency or energy use. The molecular composition, phase behaviour and physical properties of these hydrocarbon resources are complex. Significant experimental and modeling challenges are present at all length scales. The nature and behaviours of these materials are only being discovered through creative and coordinated use of diverse and novel experimental and theoretical approaches targeting knowledge gaps at the molecular, supramolecular and bulk material length scales, and keeping track of what is really known at each step. One example, drawn from our recent work shows how to construct phase diagrams for bitumen and asphaltenes using a combination of calorimetry, rheometry, quantum mechanics, polarized light microscopy and photoacoustic infrared spectroscopy [4-6]! New data and knew knowledge are creating new opportunities in a challenged and challenging sector. 1. Fluid Phase Equilibria 264 (2008) 137-146; 2. Petroleum Science and Technology, 25 (2007) 773–790; 3. J. Chem. Eng. Data, 55 (2010) 1389-1397; 4. Fluid Phase Equilibria 268 (2008) 51–60; 5. Fluid Phase Equilibria 268 (2008), 134-141; 6. Fluid Phase Equilibria 272 (2008) 32-41.

See more of this Session: Heavy Oil and Other Challenging Resources I
See more of this Group/Topical: Topical D: Chemical Engineering in Oil and Gas Production and Other Complex Subsurface Processes

(24c) Towards a Modeling Framework for Heavy Oil Phase Behavior and Properties

Harvey Yarranton, Marco Satyro and Orlando Castellanos Diaz, Chemical and Petroleum Engineering, University of Calgary, Calgary, AB, Canada

Heavy oil and bitumen are a challenging resource to recover and process mainly due to their high viscosity. Typically, heat is used to reduce the viscosity; for example, thermal processes are used to recover in situ heavy oil and a hot water process is used to extract bitumen from mined oil sands. These methods are energy and water intensive. Solvent-based or solvent-assisted processes are a potential alternative but require the understanding of more complex phase behavior. Data, characterization methods, and modeling methodologies are all lacking for mixtures of heavy oil and solvents; hence, the development of these processes carries significant risk.

A comprehensive characterization and modeling methodology for these fluids is under development at the University of Calgary. The approach involves new characterization methods and an equation-of-state based phase behavior model supported by new property correlations. A synopsis of this approach is presented with a focus on characterization.

A key challenge in characterizing heavy oils is the large non-distillable fraction of the oil. Conventional methods are based on the distillable fraction and properties must be extrapolated over the non-distillable fraction. For heavy oils, liquid-liquid equilibrium calculations are sensitive to the characterization of the medium fraction of the oil and asphaltene precipitation is sensitive to the characterization of the heavy fraction. Therefore, phase behavior predictions require a more accurate fluid characterization than can be obtained from simple property extrapolations originally developed for distillation based methods.

An alternative to extrapolation is to more rigorously characterize the non-distillable fraction of heavy oils. The idea is to fractionate the non-distillable portion of the oil using solvent-based methods and to measure properties of these fractions including vapour pressure, molar mass, density, and heat capacity. This data will be combined with GC, distillation, SimDist, and possibly PNA analysis to create a more complete representation of the oil suitable for modeling with equations-of-state.

All of the measurements except for vapour pressure are straightforward. A deep vacuum apparatus was designed to measure the vapour pressures of low volatility heavy fractions and possible to aid in the fractionation of the lighter end of the non-distillable cut. Preliminary characterization results using this approach are presented and the sensitivity of VLLE predictions to the characterization is examined.

See more of this Session: Heavy Oil and Other Challenging Resources I
See more of this Group/Topical: Topical D: Chemical Engineering in Oil and Gas Production and Other

(24d) Measurement and Modeling of Phase Equilibrium in Heavy Oil / Propane Systems

Francesca Galimberti[1], Nicolas von Solms[2], Wei Yan[2], Birgit Elkjær Ascanius[2] and Erling H. Stenby[2], (1)Eni Spa, div.R&M, San Donato Milanese (MI), Italy, (2)Department of Chemical and Biochemical Engineering, Technical University of Denmark, Kgs. Lyngby, Denmark

In order to model phase equilibrium of heavy oil with propane, several experimental measurements were undertaken using four samples from two different heavy crude oils, two as natural feeds and two as products of previous treatments.

The phase equilibrium experiments were performed in a high pressure PVT cell consisting of heavy oil/ propane mixtures at a range of temperatures, pressures and solvent/oil ratios. The densities of the two resulting phases were measured at experimental conditions. The oil and the two phases resulting from the experiments (light and heavy) were further separated by SARA characterization (they were divided into Saturates, Aromatics, Resins and three asphaltenes fractions) and the mass, molecular weight, refractive index and densities of each of the six resulting SARA fractions were measured. These values were used in order to obtain critical properties of the fractions for phase equilibrium modeling with an equation of state (Peng-Robinson). About 70 equilibrium experiments were performed altogether.

An extensive modeling effort was also carried out using the Peng-Robinson equation of state, combined with an analysis of several correlations for critical properties, acentric factors and binary interaction parameters. The model gave results for oil yield, volume fraction of the phases as well as densities of the phases. Agreement was good in most cases, despite the complexity of the systems and the errors inherent in the various characterization methods.

See more of this Session: Heavy Oil and Other Challenging Resources I
See more of this Group/Topical: Topical D: Chemical Engineering in Oil and Gas Production and Other Complex Subsurface Processes

(24e) The Compositional Continuum of Petroleum: Detailed Molecular Characterization of Heavy Crude Oils and Asphaltenes by Ultrahigh Resolution FT-ICR Mass Spectrometry

Amy M. McKenna[1], Ryan P. Rodgers[1], Christopher L. Hendrickson[1], Gregory T. Blakney[1], Joshua T. Savory[1], Nathan T. Kaiser[1], Mmilili M. Mapolelo[1], Alan G. Marshall[2], Winston K. Robbins[3], Brandie M. Ehrmann[2] and Chang S. Hsu[4], (1)Ion Cyclotron Resonance Program, National High Magnetic Field Laboratory, Tallahassee, FL, (2)Ion Cyclotron Resonance Program, Florida State University, Tallahassee, FL, (3)Consultant, Brunswick, ME, (4)Ion Cyclotron Resonance Program, Consultant, Brunswick, ME

The world's current and future dependence on fossil fuels will shift the types of materials slated for production. Light, sweet crude will diminish as their reserves approach exhaustion and the production of heavier, more heteroatom-rich crudes will increase to meet world demand. As petroleum producers increase oil production to meet the world's growing energy needs, they will inevitably access an ever-widening quality range of petroleum materials (i.e., bitumen and heavy oils) and tap into deeper and deeper offshore reservoirs. Canada has continuously ramped up oil sand mining in Alberta over the past two decades to surpass the Middle East as the primary supplier to the United States. Richer in heteroatom content and present in severe environments (temperature and pressure), these materials pose major flow assurance challenges. Thus, detailed information on oil composition is paramount in understanding its reactivity and production behavior. Oil companies sell molecules and characterization at the molecular level is vital to determine how an oil will behave during production. However, the inherent complexity of crude oil makes it difficult to provide compositional information on even the most likely culprits in common production and refining problems. In fact, many of the most common characterization techniques are based on bulk properties or solubility. Fourier transform ion cyclotron resonance mass spectrometry (FT-ICR MS) allows detailed characterization of complex petroleum samples at the level of elemental composition assignment. Ultrahigh resolution (450,000-700,000 at m/z 500) enables identification of isobaric species that differ in mass by 3 milliDalton or less, and high mass accuracy (better than 300 ppb mass error) allows for unambiguous molecular formula assignment to each of more than ten to forty thousand peaks in each mass spectrum. Thus, it is now possible to identify, sort and monitor simultaneously thousands of elemental compositions as a function of boiling point as well as provide a comprehensive analysis of non-boiling species.

Proposed over 20 years ago, the Boduszynski model describes the evolution of petroleum composition as a function of atmospheric equivalent boiling point [1]. Based on his model, Boduszynski concluded that "most of petroleum components do not exceed a molecular weight of about 200." He acknowledged that the results are controversial, "These findings are significant because of the existing controversy over whether there is an appreciable concentration of molecules in petroleum having molecular weights greater than 2000 Da. Data show there is not." However, a definitive proof of Boduszynski's model requires direct, complete compositional characterization of complex distillate cuts unavailable at that time. If substantiated, the Boduszynski model would impose strict limits on molecular weight distribution for both distillable and nondistillables petroleum fractions that contradict many previously published assertions about petroleum molecular weight and composition.

Here, we present our cumulative efforts in heavy oil and asphaltene characterization by (+/-) electrospray and atmospheric pressure photoionization FT-ICR mass spectrometry; summarize class, type, and molecular weight trends for distillable/nondistillable petroleum species; and compare results to proposed asphaltene structural models. In addition, structural characterization of nickel and vanadyl porphyrins on from an unfractionated asphaltene will be presented in a single mass spectrum for the first time. Detailed characterization of an Arabian heavy crude oil distillation series defines maltene and asphaltene compositional space (aromaticity and carbon number), and calculation of the hydrogen to carbon ratio (H:C) bounds the distillable space for all identified classes, revealing that projection of the compositional continuum to high carbon number cannot account for bulk H:C ratios. Chromatographic fractionation of crude oils from various geographic regions with varying total acid number and heteroatom content followed by subsequent mass spectral characterization reveals compositional differences and provides further insight into the separation and characterization of nonpolar and polar species in crude oils. Characterization of entrained species in asphaltene aggregates will be discussed. Compositional information for distillables and nondistillables and its implications for asphaltene aggregation will be discussed in detail.

References [1] Boduszynski, M. M.; Altgelt, K. H., Composition and Analysis of Heavy Petroleum Fractions. CRC Press: New York, NY 1994.

Keywords: APPI, petroleum, Fourier Transform, Ion Cyclotron Resonance, ICR, FT-ICR, FTMS

See more of this Session: Heavy Oil and Other Challenging Resources I
See more of this Group/Topical: Topical D: Chemical Engineering in Oil and Gas Production and Other Complex Subsurface Processes

(24f) Understanding Asphaltene Aggregation and Deposition: A Necessity for the Future of Oil Research

Michael P. Hoepfner, Tabish Maqbool, Nasim Haji Akbari Balou and H. Scott Fogler, Chemical Engineering, The University of Michigan, Ann Arbor, MI

Heavy oil is an extensive but challenging source of energy. Many difficulties currently exist with traditional crude oil reserves and working with heavy oils only complicates the extraction, transportation and refining processes. This presentation will highlight why fundamental chemical engineering research on petroleum is tantamount to the efficient handling of heavy crude oils. Central to this area is the study of asphaltenes due to their large fraction in heavy oils and their strong tendency to aggregate and deposit with changes in temperature, pressure or composition. Our research group has recently shown that asphaltene aggregation occurs under conditions previously identified as stable, further complicating the understanding of an already bewildering fraction of oil [1]. We have also shown that in addition to aggregation occurring at low degrees of destabilization, asphaltene deposition also occurs under such conditions [2]. These recent findings question the validity of both previous experimental studies and many of the thermodynamic models currently used to predict asphaltene stability. Fundamental research on asphaltenes can reveal the true mechanism(s) for asphaltene stability that can be directly applied to the recovery, transportation and processing of heavy crude oil. The industrial implications of asphaltene stability will also be discussed.

[1] T. Maqbool, A.T. Balgoa, and H.S. Fogler, "Revisiting Asphaltene Precipitation from Crude Oils: A Case of Neglected Kinetic Effects," Energy & Fuels, vol. 23, Jul. 2009, pp. 3681-3686.

[2] M. Hoepfner, T. Maqbool, and H.S. Fogler, "Effects of Precipitant Concentration on Asphaltene Deposition," AIChE National Meeting, Nashville, TN, 2009.

See more of this Session: Heavy Oil and Other Challenging Resources I
See more of this Group/Topical: Topical D: Chemical Engineering in Oil and Gas Production and Other Complex Subsurface Processes

(99a) Barriers and Challenges to the Selective Removal of Vanadium From the Asphaltene Fraction

Greg P. Dechaine and Murray R. Gray, Chemical and Materials Engineering, University of Alberta, Edmonton, AB, Canada

Most crude oils contain traces of vanadyl porphyrins within their asphaltene fraction. Although these metals are only present in trace quantities, they have a significant detrimental impact on crude oil processing units; therefore, their selective removal is highly desirable. The fact that the majority of the vanadium is contained within the highly aromatic, highly polar asphaltene fraction also poses additional roadblocks to their selective removal. This fraction has been shown to associate/aggregate significantly in most (if not all) solvents. Recent work on the nature and possible mechanisms of the molecular association of asphaltenes in solution can be extended to help elucidate the molecular interactions occurring between asphaltenes and metalloporphyrins, and hence the nature of the inclusion of metalloporphyrins within the asphaltene fraction. Solubility measurements with simple vanadyl porphyrins showed that they fit the operational definition for asphaltenes, exhibiting negligible solubility in n-heptane and orders of magnitude higher solubility in toluene. Further studies were also carried out using a stirred diffusion cell. These experiments determined that the size of the asphaltene aggregates in toluene at 25°C and 1 g/L were between 5 and 9 nm. These aggregated structures were extremely stable, remaining intact for periods of >7 days. Neither an increase in temperature nor a decrease in concentration (to 0.1 g/L) reduces the size of these structures below 5 nm, and the origin of the asphaltenes (Athabasca, Safaniya, Venezuela) did not have a significant impact. A significant amount (>85%) of the native petroporphyrins present in the asphaltene fraction remain bound within the above aggregated structures and hence are not susceptible to selective removal. The native petroporphyrins that are free to diffuse appear to be larger than model vanadyl porphyrins as indicated by pore hindrance effects. Neither an increase in temperature nor a decrease in concentration results in an increase in the amount of petroporphyrins available for diffusion. The selective separation of metalloporphyrins is clearly hampered by gaps in the basic understanding of the aggregation behavior of the asphaltene fraction, as well as gaps in understanding metalloporphyrin properties and behavior in solution.

See more of this Session: Heavy Oil and Other Challenging Resources II
See more of this Group/Topical: Topical D: Chemical Engineering in Oil and Gas Production and Other Complex Subsurface Processes

AIChE Conference, Annual Meeting 2010

SOLUBILITY AND MASS TRANSFER PARAMETERS OF HYDROGEN IN VACUUM RESIDUES UNDER HYDROCONVERSION CONDITIONS

Mariela Sanoja, Laurent Sehabiague, Yannick Heintz, Rachid Oukaci* and Badie I. Morsi
Chemical and Petroleum Engineering Department
University of Pittsburgh, Pittsburgh, PA 15261

**Energy Technology Partners, LLC*
Pittsburgh, PA 15238

The volumetric liquid-side mass transfer coefficient (k_La) and equilibrium solubility (C^*) for H_2 were measured in two different vacuum residues (Zuata and Merey-Mesa) using one-liter agitated autoclave operating in a Gas-Inducing mode under the typical conditions of hydroconversion of heavy residues. The effects of operating variables, including pressure (up to 55 bar), temperature (up to 300 °C), mixing speed (up to 2000 rpm) and solid concentration (up to 40 wt%), on k_La and C^* were investigated using the Central Composite Statistical Design and Analysis technique. The transient physical gas absorption method and a modified Peng-Robinson Equation-of-State were used to obtain k_La and C^*, respectively.

Under the operating conditions investigated, k_La values for H_2 were found to increase with increasing pressure, temperature and mixing speed, and to decrease with increasing solid loading. Also, the equilibrium solubilities for H_2 in the two vacuum residues appeared to increase with pressure at constant temperature and increase with temperature at constant pressure. These data are indispensable for proper design and scaleup of slurry reactors heavily employed in hydro-processing of vacuum residues.

(99c) On Specific Partial Molar Volume of Asphaltenes in Organic Liquids and Their Solubility Parameters

Kasra Nikooyeh and John M. Shaw, Chemical and Materials Engineering, University of Alberta, Edmonton, AB, Canada

Similarity of Hildebrand solubility parameter is frequently used in petroleum science, as a measure of compatibility of constituents and is commonly used for interpreting and correlating asphaltene-diluent interactions. Specific partial molar volume at near infinite dilution is a sensitive measure of solute-solvent interactions derived from high precision density measurements of dilute mixtures. It equals the partial molar volume of a solute divided by its molar mass, but is obtained directly from the density measurements without invoking a molar mass for the solute. Such a measurement is appropriate here because the mean molar mass of asphaltenes is not known a priori. In this contribution, the specific partial molar volumes of Athabasca and Maya heptane asphaltenes at near infinite dilution are reported for organic liquids over the temperature range 20 - 80 C. The organic liquids comprise pure compounds and binary mixtures including: heptane, toluene, 1-methylnaphthalene, quinoline, anisole, 2,6-lutidine, pyridine, methylene chloride, tetrahydrofuran, decane). At 20 C, these organic liquids possess solubility parameters ranging from 15 MPa0.5 (heptane) to 22 MPa0.5 (quinoline and methylene chloride). Surprisingly, the solubility parameter of the organic liquids, in this range, does not appear to correlate with the specific partial molar volume of these asphaltenes at near infinite dilution. Estimation of the solubility parameter of asphaltenes is discussed.

See more of this Session: Heavy Oil and Other Challenging Resources II
See more of this Group/Topical: Topical D: Chemical Engineering in Oil and Gas Production and Other Complex Subsurface Processes

(99d) Upgrading of Athabasca Oilsands Bitumen by Low Temperature Natural Zeolite Cracking

A.S.M. Junaid[1], W. Wang[2], C. Street[2], M. Rahman[2], W. McCaffrey[2] and S. Kuznicki[2], (1)Chemical and Materials Engineering, University of Alberta, Edmonton, AB, Canada, (2)University of Alberta, Edmonton, AB, Canada

Synthesized or acidified minerals, particularly clays, have been found to effectively break heavy petroleum resources such as gas oil. However, the application of natural zeolites (NZs) as cracking catalysts and/or adsorbents for oilsands bitumen is a novel approach. Canadian oilsands bitumen, the majority of which is in Alberta, represents approximately 15% of the proven global crude oil resources, and is an extremely valuable source of energy for North America. However, the density, viscosity, contaminant content, and residue fraction of the bitumen recovered from these sands are much higher than those of conventional crude oil due to the presence of large molecules such as asphaltenes. Our studies show that acidified NZs such as clinoptilolite from Saint Clouds, New Mexico and Ca-chabazite from Bowie, Arizona can be developed as very effective reagents for cracking of oilsands bitumen. Low-cost NZs, particularly clinoptilolites, are present globally in large natural deposits. NZ-cracked oilsands bitumen products are highly recoverable (up to ~84%) using light hydrocarbons such as pentane, which indicates substantial conversion of heavier fractions to lighter components. Boiling point distribution and viscosity measurements of the products demonstrate that NZ catalyzed cracking results in much less viscous low boiling point products compared to those produced from pure thermal treatment, or analogous catalytic treatment using the standard petroleum cracking agent, zeolite Y. An additional benefit of using NZs for oilsands bitumen processing is that they function as strong adsorbents to facilitate removal of contaminant heteroatoms such as vanadium (up to ~80%) and nitrogen (up to ~38%). Based on our findings we envision a NZ-based alternative to traditional primary upgrading processes for bitumen, that operates at lower temperatures, improves the transportability of bitumen, and simultaneously removes contaminants, reducing processing steps. This study investigates the efficacy of NZ catalyzed cracking for conversion of heavy fractions of Athabasca oilsands bitumen. Boiling point distribution, average molecular weight and viscosity of the catalytically cracked samples are compared with baseline values for raw and thermally cracked bitumen samples.

See more of this Session: Heavy Oil and Other Challenging Resources II
See more of this Group/Topical: Topical D: Chemical Engineering in Oil and Gas Production and Other Complex Subsurface Processes

SUPERCRITICAL WATER TO UPGRADE PETROLEUM FEEDSTOCK

Ki-Hyouk Choi, Ali M. Al-Somali, Mohammad F. Aljishi, Joo-Hyeong Lee, Mohammed R. Al-Dossary, Ashok K. Punetha, Bader M. Al-Otaibi, and Mohammad Z. Al-Bahhar
Research & Development Center, Saudi Aramco, Dhahran, Saudi Arabia

Introduction

Depleting good-quality crude oil necessitates the reliance on the abundant resources of non-conventional hydrocarbons and heavy & sour crude oils to meet the steadily increasing oil demand while supplying "clean" fuel to the market. However, these resources require major treatments since they are associated with low API gravity and high sulfur content, which present many challenges for refineries to manage these crude oils until final products. Alongside these obstacles, environmental regulations are getting more stringent to minimize air pollution. Growing cleaner alternative energy sources, biofuels for example, exert pressure on crude oil products, which are considered environmentally harmful. This presents an additional challenge in upgrading heavy and sour hydrocarbons. One of the most important processes for refining petroleum based fuels to meet the regulations on transportation fuels is desulfurization.

Available Technologies

At present, there are two major existing technologies to upgrade heavy and sour oil: carbon rejection and hydrogen addition [1]. Carbon rejection results in breaking down hydrocarbons into smaller chains. However, it results in the formation of coke, hence, a product loss. The product of carbon rejection is often unstable and requires further treatment. In general, carbon rejection is an inefficient process that does not lead to substantial results in upgrading heavy and sour oil. These factors make carbon rejection an unattractive option. The alternative, hydrogen addition, make use of hydrogen and a catalyst to upgrade heavy and sour oil. Coke formation is mostly avoided in this process. Although this process is preferred over carbon rejection, the use of hydrogen and a catalyst binds the process to a high cost which presents economical limitations. These factors call for innovative ideas to upgrade heavy and sour oil. A lot of effort is done to obtain new processes as alternative technologies to the existing ones for upgrading and desulfurization of heavy and sour oil. Different non conventional methods have been proposed and evaluated. For example, ultrasonic irradiation was reported to show certain performance in upgrading and desulfurizing heavy oil. A method is also discussed and introduced in this paper which is using supercritical water , which acts as an organic solvent at the critical point and pressure, to reduce the amount of sulfur in the heavy and sour oil.

Properties of SCW

Hydrothermal treatment of heavy and sour oil through supercritical water (SCW) is a viable option. SCW, which is water at temperatures over 374 °C and a pressure over 218 atm, exhibits distinctive chemical and physical behavior [2]. At supercritical conditions, hydrogen bonding in water is substantially weakened, which results in a significant increase in water ionic properties and a decrease in its dielectric constant. The ionic product of water could increase as much as 3 orders of magnitude, which results in an excess of $[H^+]$ or $[OH^-]$. This makes reactions involving SCW, whether acidic or basic, self-catalyzed. At supercritical conditions, water density is substantially reduced as well as viscosity, which minimizes the barrier for rotational and translational motion, enhancing water transport properties as a reaction stimulating solvent. Through cage effect, which stabilizes free radicals, coke formation is avoided. In addition, supercritical water stabilizes the transition state by modifying the geometry, easing the path toward upgraded heavy and sour oil. These unique chemical and physical behaviors could be utilized to upgrade heavy and sour oil. This is achieved through homolytic cleavage of covalent bonds which produces free radicals. Supercritical water-treatment of heavy oil achieves many goals at once; significant reduction in sulfur content, increase in API gravity, and reduction in asphaltene and metal content. The graph below shows the relationship between the temperature and dielectric constant at different pressures. Dielectric constant of supercritical water is very near to that of dichloromethane(DCM) and toluene[3].

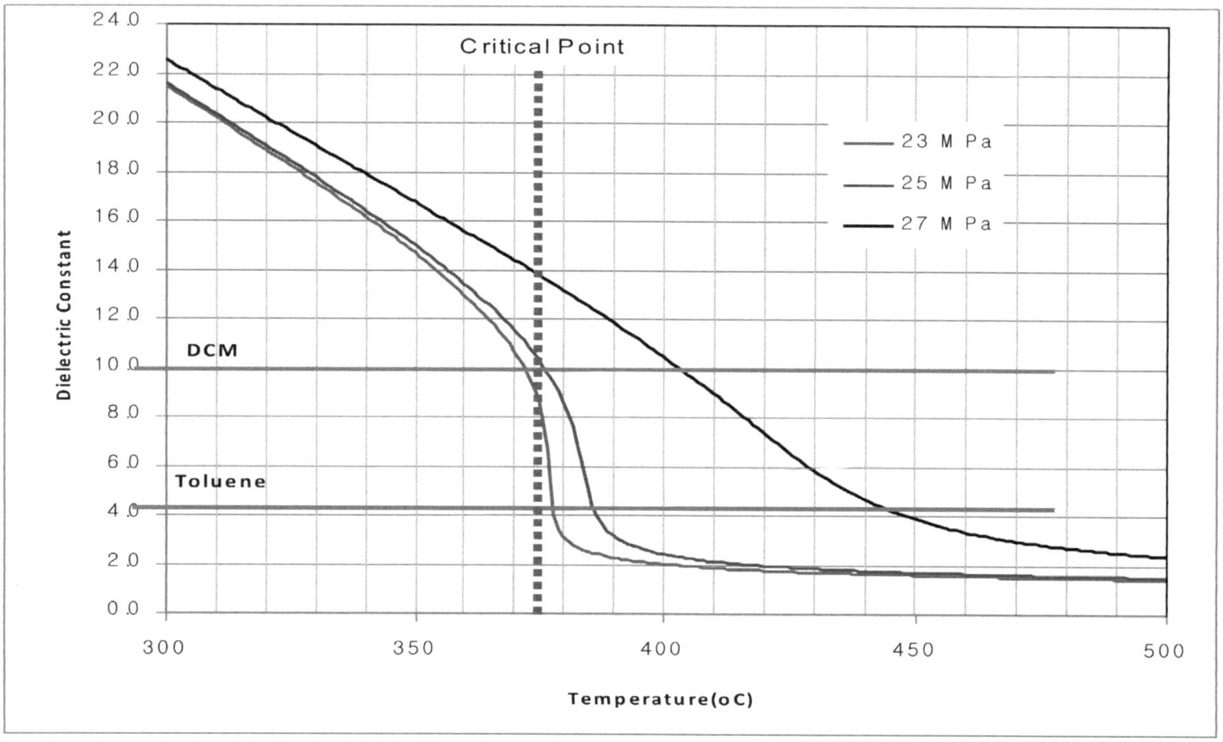

Fig. 1. Dieletric constant of supercritical water.

Experimental

A series of model compounds was tested to elucidate probable mechanism of reactions occurring in supercritical water environment. As a representative sulfur compounds included in heavy oil, thiophene, benzothiophene, and hexyl sulfide were selected for testing. Also, tetralin, decalin, dodecane, and toluene were used to simulate matrix of heavy oil derived from petroleum. Content of sulfur compounds was adjusted to between 0.5 – 2.2 wt% sulfur. All chemicals used for model compound study were obtained from Aldrich and had high purity. **Table 1** summarized some properties of heavy oil used for this study.

Table 1. Properties of Heavy Oil

Property	**Measured Value**	
Gravity	27.0 API Degree	
Total Sulfur Content	3.0 wt% Sulfur	
Distillation Property (GC-SIMDIS)	5 vol%	92 °C
	10 vol%	152 °C
	20 vol%	249 °C
	30 vol%	333 °C
	40 vol%	424 °C
	50 vol%	515 °C
	60 vol%	628 °C
Kinematic Viscosity(SUS)	38 SUS at 100 F	

Testing was conducted by using a reactor made of stainless steel tube reactor (0.25 inch or 0.5625 inch O.D., 0.083 inch or 0.188 inch I.D.) maintained at 380 - 485 °C by electric heater. Deionized water and oil were pumped into the reactor by high pressure metering pumps through separate lines which were heated to certain temperatures. The oil feed including model compounds, was not pre-heated to avoid conversion before combing with supercritical water. Water to oil ratio was kept at 1 or 2 by vol/vol. Residence time of feed mixture in the reactor was calculated by assuming that the density of model feed was the same to that of water in the reaction condition. Product stream from the reactor was cooled to room temperature before depressurization to atmospheric pressure. The liquid phase product is collected during certain periods to measure total liquid mass balance which was almost 100 wt %. The gaseous product was analyzed by on-line mass spectrometer to obtain qualitative composition. No solid deposit was observed in the reactor after the reaction. GC-MS(Agilent 7890 and MSD), GC-FID(Agilent 6890), and GC-SCD(Agilent 6890 coupled with Sievers 355) were used to identify and quantify product molecules. Total sulfur content and SIMDIS were measured by relevant ASTM methods. Also, 1H and 13C-NMR were taken to analyze change of hydrocarbons during supercritical water treatment.

Results and Discussion

Upgraded oil results indicate that a 15-30% reduction in sulfur content was achieved while increasing API gravity without generating coke at the reaction conditions of 380-450 °C and 3,200 – 3,600 psig.. **Table 2** shows the amount of sulfur removed during the tests.

Table 2. Properties of Feed and Product Oil.

	Feed	Product	Conversion
Total Sulfur	1.27%	1.16%	8.9%
Thiophene	0.67%	0.59%	11.5%
Benzothiophene	0.31%	0.30%	5.3%
Hexyl Sulfide	0.29%	0.28%	5.1%

Reaction temperature, pressure and residence time used for the testing of Table 2 were 413 °C, 1.53 seconds, and 3,600 psig, respectively.

Study with model compounds was conducted to investigate the effect of matrix hydrocarbon on desulfurization of thiophene. As shown in Fig. 1, dodecane showed the highest desulfurization activity than that of tetralin and decaline. Conversion of thiophene in decalin was much lower than that in tetralin. In general, decomposition of thiophene was understood to proceed through a radical mechanism.[3] The role of supercritical water in such a radical reaction has been proposed as the hydrogen donor, stabilizer, and dilution medium.[2] Impact of matrix hydrocarbon on the conversion of thiophene was understood to be related to participation of the hydrocarbon in the radical reaction network. As noted above, n-dodecane generated olefin and paraffin at the reaction conditions employed in this study while decalin and tetralin were relatively stable during reaction. From such observations, it was suggested that n-dodecane facilitated radical reactions which were responsible for thiophene conversion, by generating large amount of radicals in the reactor. In contrast, tetralin was known to stabilize radical by hydrogen donation and inhibit further decomposition or combination of the radical[4] although hydrogen donation might facilitate thiophene conversion. It must be noted that the difference of thiophene conversion was significant, what indicates the important role of matrix hydrocarbons in desulfurization.

Products from heavy oil showed the increase of hydrogen to carbon ratio as shown in Fig. 2. Increase hydrogen amount of product seems to be from hydrogen donation from water or steam reforming reaction. Products having lower sulfur content had higher hydrogen to carbon ratio. The amount of saturate and aromatics increased while the amount of asphaltene and resin has decreased by supercritical water treatment as shown in Fig. 3. Also, it was observed that the significant amount of asphaltenic sulfur was changed to maltenic sulfur which is shown in Fig. 4. The graph shows the amount sulfur transferred in the feed and in the product. Gas-phase product contained hydrogen, light hydrocarbons such as ethane, carbon dioxide and also hydrogen sulfide as in Fig. 5. Although mass spectra of Fig. 5 doesn't have quantitative information, it is obvious that some amount of hydrogen was generated in the reactor and sulfur was transformed into hydrogen sulfide. Source of hydrogen gas is not identified yet. However, steam reforming reaction can be probable source of hydrogen and also carbon dioxide

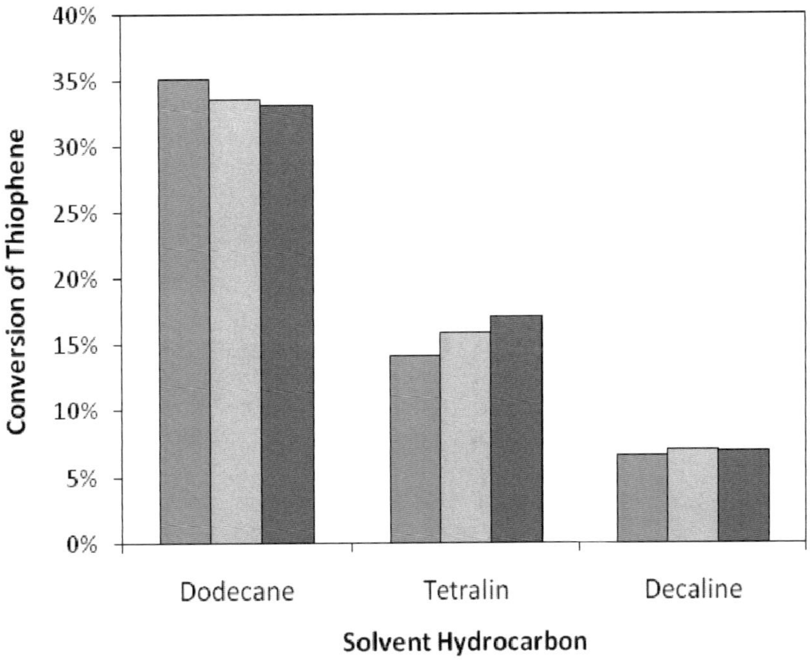

Fig. 1. Thiophene conversion in various matrix hydrocarbon. Reaction temperatures were 403 °C, 457 °C, 485 °C(from left to right). Residence time of feed mixture was in the range of 1.6 – 1.9 seconds at the operating condition.

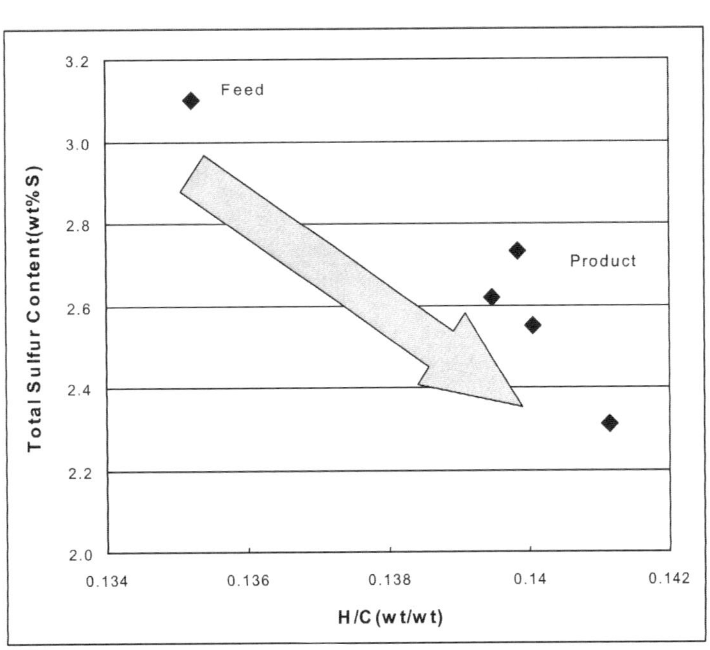

Fig. 2. Hydrogen to carbon ratio of product vs. total sulfur content of the product.

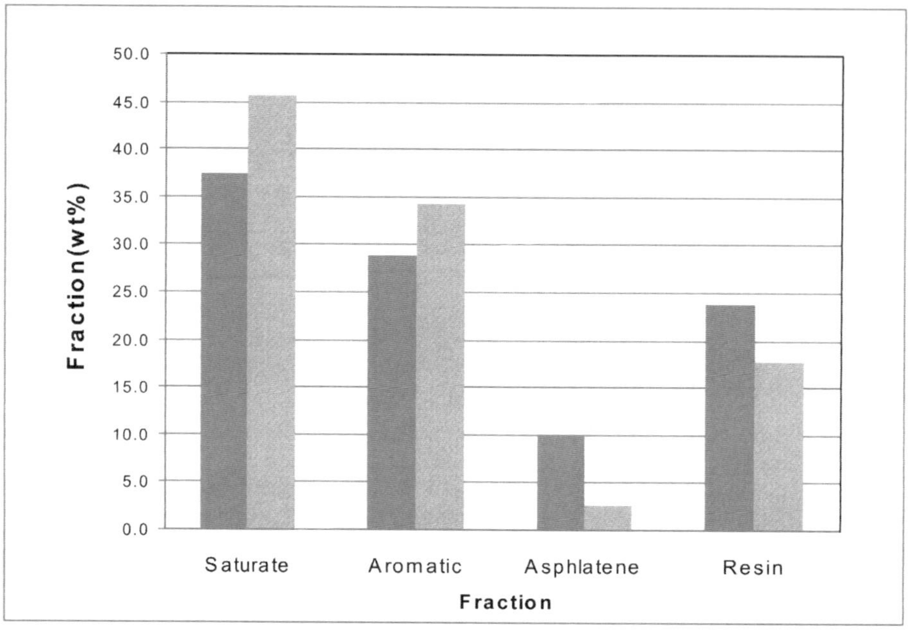

Fig. 3. Saturate, Aromatic, Asphaltene, and Resin amounts of feed(left) and product(right).

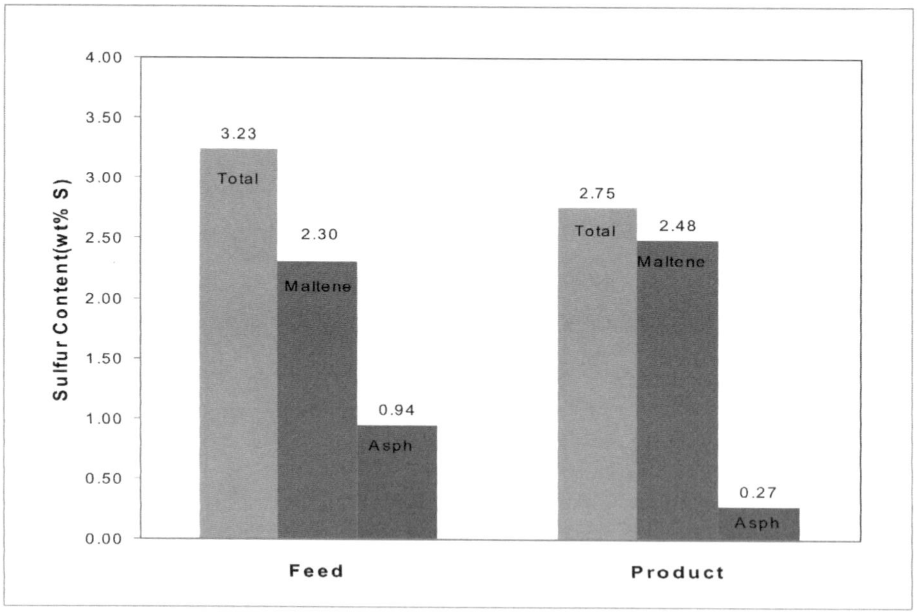

Fig. 4. Sulfur contents of Saturate, Aromatic, Asphaltene, and Resin fractions of feed and product.

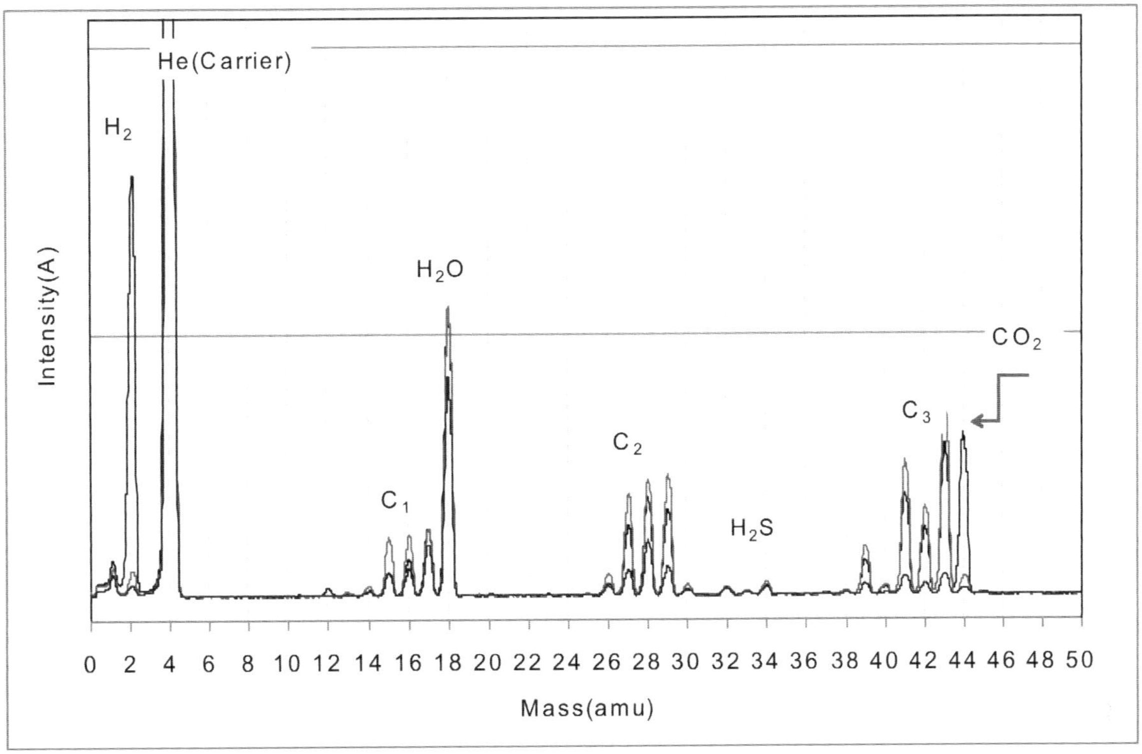

Fig. 5. Mass spectra of product gas. Black and red spectra were taken from different operating codntions. Helium was used as a carrier gas due to small amount of gas-phase product.

Conclusion

This study presents hydrothermal treatment as an innovative technology for upgrading heavy and sour oil. It discusses the unique properties of supercritical water and its role in upgrading heavy and sour oil as well as its applicability in the mid-stream operations sector. Upgrading reactions occurring in the supercritical water conditions were related to radical-mediated network and strongly influenced by co-present hydrocarbons. This process takes advantage of water, which is abundant in upstream operations, in an environmentally friendly manner and self-catalyzed conditions. It also eliminates the need of hydrogen, thus overcoming the economic limitation of hydrogen addition, and avoids product loss through coke formation. In spite of very promising results shown in this study, many technical hallenges still exist to improve the quality of upgraded heavy and sour oil. Challenges, such as corrosion, high energy requirement, and reactor design need to be resolved. Although corrosion is very notorious for supercritical water process, a Japanese group suggested that conventional stainless steel due to absence of oxygen and halogen in feedstock [4]. More comprehensive experiments should be carried to determine optimized conditions for oil-to-water ratio, reaction temperature, and residence time.

Acknowledgement.

Authors are grateful to Dr. Esam Z. Hamad for fruitful discussion.

References

(1) I. V. Babich and J. A. Moulijn, *Fuel*, **82**, 607(2003).
(2) N. Akiya and P. E. Savage, *Chem. Rev.*, **102**, 2725(2002).
(3) M. Uematsu and E. U. Franck, *J. Phys. Chem. Ref. Data*, **9**, 1291(1980).
(4) R. Fujisawa, M. Sakaihara, Y. Miyashita, W. Saiki, and R. Mori, *J. Japan Inst. Energy*, **88**, 180(2009).

.

(99f) Separation of Heavy Oil Emulsions

Jayant P. Rane, Energy Institute, Department of Chemical Engineering, City College of New York, New York, NY, David Harbottle, The Energy Institute, The City College of The City University of New York, New York, NY and Sanjoy Banerjee, Chemical Engineering, Energy Institute, City College of New York, New York, NY

As the conventional light oil is replenishing, the focus has been shifted to the recovery of heavy oil. Almost half of the world's reserves are classified as heavy or extra-heavy. For better understanding of the heavy oil recovery process, it is important to understand the gravity separation process using heavy oil system. Heavy oils have density typically in the range of 10-30 API causing much lower separation rate than conventional crude oil due to very low density difference between two immiscible liquids i.e. oil and water. To enhance the separation rate and achieve similar processing throughputs as observed in conventional crude oil processing, water droplet-droplet coalescence and droplet-homophase coalescence should be maximised. Thus it is important to understand droplet-droplet and droplet-homophase interactions. In a typical industrial process, much of the droplet-droplet coalescence occurs within the process pipeline leading into the gravity settler. However, there is still great uncertainty regarding droplet-droplet coalescence during separation. Aim of the current study is to understand these coalescence mechanisms in the continuous gravity settler and also, to study the droplet-droplet and droplet-homophase interactions independently in the microfluidic cell.

Laboratory scale vertical gravity settlers (height: 1m, diameter, 0.075m) have been constructed to study the coalescence mechanics during separation. To reduce the number of variables that are to be considered, model emulsions have been developed using mineral oil and de-ionised water, stabilised by Span 80. Also, some experiments have been carried out using asphaltenes instead of SPAN80. For the heavy-oil emulsion investigation, mineral oil is premixed with Halocarbon oil (ρ = 1.95g/cc) to increase the oil density to 0.94g/cc. To allow for measurements under steady state conditions a closed loop set-up has been utilised. The droplet size distribution and the emulsion density are measured along the length of the column at the steady state. From the drop size distribution it is possible to track the changes in size as a function of position. Such information is important when trying to understand coalescence mechanics. Through a combination of both measurements, and measurement of the water retrieval flow rate, a coalescence rate at the emulsion-water interface can be inferred. The experimental program considers both the separation of light-oil and heavy-oil systems as a function of surfactant concentration. At steady state, water outlet in the gravity settler corresponds to droplet-homophase coalescence rate. It is found that, this water outlet decreases with increase in surfactant concentration.

Microfluidic setup is used to generate water droplets in the range 20-100 microns in oil using flow-focusing geometry made up of PDMS using soft lithography. In order to understand how droplets interact with each other, droplets are brought into contact with the range of different velocities and surfactant concentrations (Capillary number). Nature of interaction and interaction time is recorded to get a broad spectrum of data. Such data is relevant in the sedimentation zone of the gravity settler where droplet-droplet interactions are predominant. For droplet-homophase interactions, generated droplets are brought in contact with the pool of water homophase at T-junction and droplets are stacked at particular pressure over the pool of water homophase. Such kind of interactions is observed near the packed bed-homophase interphase in the gravity settler.

See more of this Session: Heavy Oil and Other Challenging Resources II
See more of this Group/Topical: Topical D: Chemical Engineering in Oil and Gas Production and Other Complex Subsurface Processes

(149a) Reducing the Impacts of Resource Temperature Degradation and Varying Ambient Temperature On Air-Cooled Binary Plant Performance

Daniel S. Wendt, Advanced Process & Decision Systems, Idaho National Laboratory, Idaho Falls, ID and Gregory L. Mines, Biofuels & Renewable Energy Technology, Idaho National Laboratory, Idaho Falls, ID

As the geothermal industry moves to use geothermal resources that are more expensive to develop, there will be increased incentive to use more efficient power plants. Because of increasing demand on finite supplies of water, this next generation of more efficient plants will likely need to reject heat sensibly to the ambient (air-cooling). Though they use no water, air-cooled systems have higher capital costs, reduced power output (rejection at a higher temperature), lower power sales due to higher parasitics (fan power), and have greater variability in power output (variation in the dry-bulb temperature). These effects are compounded as the resource temperature decreases.

Several currently available technologies have been studied to evaluate their potential to cost-effectively increase plant overall efficiency and total power generated during the plant operational life. These technologies include recuperation, in which turbine exhaust superheat is used to preheat the condensed working fluid prior to heat exchange with the geothermal fluid; reheat, where an intermediate heat exchange between the working fluid and geothermal fluid takes place during the expansion of the working fluid in the turbine; non-consumptive use of EGS makeup water to supplement the heat rejection; and substitution of working fluids at an intermediate point in the geothermal plant operational life.

Aspen Plus process models were developed to calculate optimal equipment sizing for maximizing the net power output of various process configurations in which geothermal resource temperature, geothermal fluid flow rate, ambient temperature, working fluid composition, heat exchanger pinch points, turbine and pump efficiencies, and process equipment and piping pressure drops were specified. A minimum geothermal fluid outlet temperature constraint was set to mitigate concerns of mineral precipitation on the heat exchanger surfaces. Two scenarios modeled were a baseline geothermal process employing the Organic Rankine Cycle (ORC) and a modified geothermal process design employing recuperation, reheat, and makeup water heat rejection in the Organic Rankine Cycle.

Following completion of the "plant design" modeling described above, a process modeling analysis was performed to evaluate the effect of hourly ambient temperature variation and declining resource temperature over the operational life of the two power plant designs. The equipment specifications from the baseline and modified geothermal plant process designs were fixed and correlations were incorporated into the process models to adjust heat transfer coefficients, pump and turbine efficiencies, and pressure drops in order to evaluate process performance as geothermal resource and ambient temperatures varied. The capability to vary turbine nozzle geometry was included in all simulations to provide an additional degree of freedom in establishing optimal process conditions. In all cases evaluated, the geothermal fluid flow rate was assumed constant.

A comparison of the baseline geothermal plant and modified geothermal plant capital costs and total power generated during the plant operational life was then performed for a scenario with a defined resource temperature decline and annual temperature profile. The methodologies used to select the plant design conditions, including ambient design temperature, working fluid selection, and turbine nozzle area for a declining resource temperature are discussed.

See more of this Session: Geothermal Engineering
See more of this Group/Topical: Topical D: Chemical Engineering in Oil and Gas Production and Other Complex Subsurface Processes

A Downhole Fluorimeter for Measuring Flow Processes in Geothermal and EGS Wellbores

Michael Mella[1], Steven Olsen[2], Peter Rose[1], Daniel Bour[3]Susan Petty[3]and Joel Harris[4]

[1]Energy and Geoscience Institute at the University of Utah · 423 Wakara Way Suite 300, Salt Lake City, UT 84108

[2]Red Rock Analog Design · 1218 Emerson Ave, Salt Lake City, UT 84108

[3]Alta Rock Energy Inc · 7900 E Green Lake Drive N Suite 202, Seattle, WA 98103

[4]Department of Chemistry, University of Utah

Keywords

Borehole Fluorimeter, Borehole Flowmeter, Fracture Characterization, EGS, Geothermal

Abstract

The conventional approach to measuring flow downhole in geothermal or EGS wellbores is through the use of a mechanical spinner tool, which makes use of a rotating paddle suspended and lowered (or raised) through the wellbore on a cable. But, such tools are notorious for failing-especially at high flow rates and at high temperatures. Likewise, non-uniform wellbore diameters greatly complicate spinner-tool data interpretation. An alternative approach to measuring flow rates in a geothermal or EGS wellbore is described. Making use of conventional tracer dilution methods, the tool is designed to work under both laminar and turbulent flow regimes. Volumetric flow rate measurement will allow the tool to accurately measure flow even through washout regions where borehole diameter varies. The tool is designed to be deployed on a single-conductor wireline and will be heat shielded. An optical fiber bundle will serve both to deliver the tool's excitation light signal and to return the emission signal from the fluorescent dye that will be introduced from a pump attached to the end of the tool. Temperature and pressure measurements will also be made using Resistance Temperature Detectors (RTD) and pressure transducers.

Introduction

Downhole temperature, pressure and flow rates are essential pieces of information for any geothermal field. Wire-line tools transmit data to the surface giving real-time readouts with depth. From the data logs fluid entry zones and temperature profiles of the wellbore can be obtained. Flow rates can be calculated and fluid entry zones can be interpreted in combination with temperature data. Temperature-Pressure-Spinner (TPS) tools are the current industry standard in obtaining this valuable information. TPS tools are fairly simple and straight forward instruments to understand and operate and can be manufactured to operate at high temperatures for extended periods of time (Davarzani 1988). When combined with caliper log data in combination with temperature and pressure data, TPS log data can be used to calculate volumetric flow rates.

As important as TPS tools are and how much valuable data they provide, there are limitations of TPS tools. TPS reliability relies on the velocity of water to turn a mechanical device on the tool. Flow rates that represent the two extremes in geothermal systems pose problems for TPS tools. Spinner tools typically need velocities in excess of 5 ft/min to turn the impeller element (Smollen 1996). Flow rates in the laminar flow range are not measurable. At very high flow velocities spinner tools may not be able to keep up with the velocity and the impeller itself may break off. The flow in between the two extremes may also require tools with differing impellers to accurately measure velocities. A second drawback to TPS tools is the inability to accurately measure flow in some open-hole or wash-out regions. If the tool is in a space of unknown diameter the velocity may not be used to accurately calculate the volumetric flow rate. The tool that is detailed in this paper is one that uses measured tracer concentration to determine flow rates.

Measuring flow through a pipe using tracing dyes dates to when Reynolds conducted his flow experiments in the late 1800's (de Nevers, 2005). Radioactive tracer tools have been employed for measuring flow in oil field injection wells (Smollen 1996). Radioactive tracers have the obvious drawback of using radioactive materials. For this reason radioactive tracing is used almost exclusively in injection wells. Most radioactive tracer tools do not measure concentration but rather are time of flight type tools that measure tracer pulses passing two different detectors.

The fluorescent tracers used for the application described in this paper need to be highly detectable and thermally stable. These compounds will additionally have to fluoresce at a wavelength where there is little interference coming from naturally occurring chemicals in the geothermal brine. Among the chief tracers being considered is fluorescein. Fluorescein has been proven to be highly detectable and thermally stable for use as a geothermal tracer (Adams et al. 1991). Other tracers that could be used include one in the family of naphthalene sulfonates (Rose et al. 2003), PTSA (Rose et al. 1998) or even quantum dot tracers (GTP 2010).
This paper describes a tool that will measure tracer concentration and calculate volumetric flow rates based on tracer dilution techniques. Using this approach, mass flow rate ($\dot{m}$) is calculated from the concentration of the tracer (X) and the rate at which the tracer is delivered (Q):

$$\dot{m}\left(\frac{m_{water}}{t}\right) = \frac{Q\left(\frac{m_{tracer}}{t}\right)}{\mathrm{X}\left(\frac{m_{tracer}}{m_{water}}\right)} \tag{1}$$

A volumetric flow rate can then be calculated using mass flow rate. Tracer dilution techniques have been successfully used for measuring flow in geothermal production wells at the surface for several years (Hirtz et al., 2001 and Lovelock, 2001). The design process of such a tool to measure and deliver tracers at geothermal conditions is the subject of the remainder of this paper.

Tool Outline

The tool will be capable of high temperature operation for extended periods of time and be able to deliver and measure water soluble tracer in the wellbore and to send information back to the surface in real-time. The tool will consist of a fiber optic at one end for detecting tracer, a tracer delivery system at the other end, and heat shielded electronics. It will be connected to a single conductor wireline that will serve as both the structural tethering element and communication element to the surface. At the surface the tool will be controlled by a computer that will be capable of sending and receiving instructions. The tool will also contain pressure and temperature measuring devices and log this information at the same time it is logging the tracer signal. A cartoon sketch is shown in Figure 1 below.

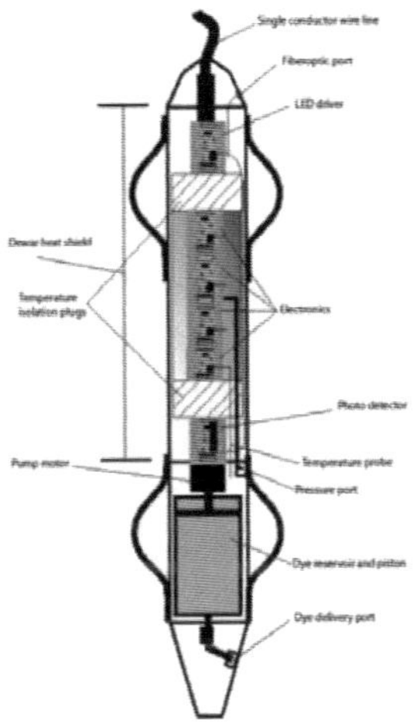

Figure 1. Sketch of the fluorimeter-flowmeter.

Flow Conditions and Tracer Mixing

Thorough mixing of the injected tracer within the wellbore is essential the accuracy of volumetric flow rates. Assuming an 8-inch well bore diameter calculations were made for several volumetric flow rates to determine the distance need to achieve 99% axial mixing in turbulent conditions. Calculation of mixing distance, L, was performed using Equation 2 listed below where f is the Fanning friction factor and is defined in equation

3.

$$L = \frac{0.56D}{\sqrt{f}} \qquad (2)$$

$$f = 0.001375\left[1 + \left(20{,}000\frac{\varepsilon}{D} + \frac{10^6}{\mathrm{Re}}\right)^{\frac{1}{3}}\right] \qquad (3)$$

Equation 3 is only valid for turbulent flow, or flows for which Re is greater than 2,300. In Equation 3, D is the pipe diameter, which was assumed to be 8 inches, and ε is the surface roughness assumed to be the roughness of commercial steel pipe. Re is the Reynolds number and is defined as

$$\mathrm{Re} = \frac{4Q}{\pi D \nu} \qquad (4)$$

where Q is the volumetric flow rate, D is the diameter and ν is the kinematic viscosity. Kinematic viscosity is a function of temperature and the fluid itself. In this case the fluid viscosity was assumed to be that of water and the viscosity at different temperatures was calculated. From these calculations a plot was constructed showing mixing length vs. volumetric flow rate. The Reynolds numbers ranged from 41,000 to $1.7\text{x}10^7$. The plot is displayed below as Figure 2.

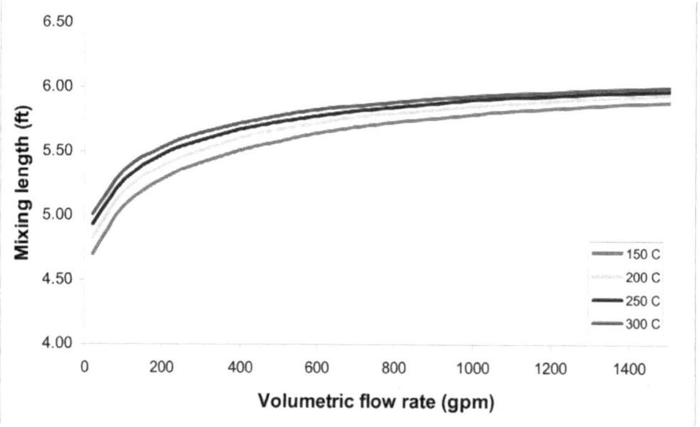

Figure 2. Mixing lengths required for complete axial mixing vs. flow rate.

From the plot one can see that the injection source and detection must be about 6 feet apart to approximate complete mixing even at very high flow rates (high velocities). The back mixing effect present in turbulent flow rates and present in most geothermal wells will ensure well mixed tracer provided the tool is longer than 6 feet.

The use of the tool at laminar flow conditions will be a challenge because of the lack of much back mixing inherent in the flow. Figure 3 below shows what tracer concentration across the wellbore might look like 8 feet from the injection source at different flow rates. As one can see from the plot, the tracer concentration would vary significantly from the borehole walls, at 1 and minus 1, to the center line.

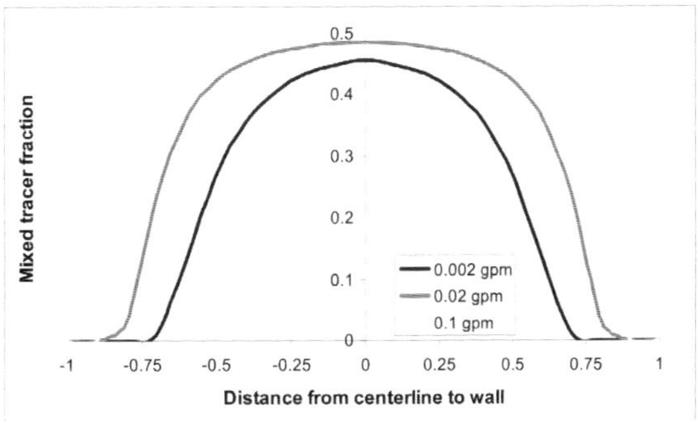

Figure 3. Tracer concentration profile across the wellbore at a distance of 8 feet from the injection point.

Operation at laminar flow regimes will require the addition of some sort of mixing device to provide for even mixing across the wellbore. Two of the ideas to help induce mixing include adding a piece that would have recirculation jets to induce back mixing or adding a mixing brush on the outside of the tool. Both of these options are depicted graphically in figure 4 below.

Figure 4. Cartoon representation of mixing devices, recirculation jets on the left and mixing brush on the right.

Temperature Effects

Perhaps the biggest challenge of deploying a tool in a geothermal wellbore is the high temperature. Using a Dewar-style heat shield in addition to a heat sink, estimates were made of the internal temperature of the tool as a function of time at geothermal conditions. With a stainless steel heat sink and an assumption of 5 watts of power dissipation within the tool, a plot of internal temperature versus time for several temperatures was produced and is shown below as Figure 5. Included in the plot are two lines that represent 85°C and 125°C. These two lines are significant because they show the temperatures at which traditional electronics (85°C) and automotive class electronics (125°C) will fail. The price increase to move from traditional to automotive grade electronics is not significant. Any electronics beyond these temperatures will need to be high temperature components that are available only at a significant increase in cost.

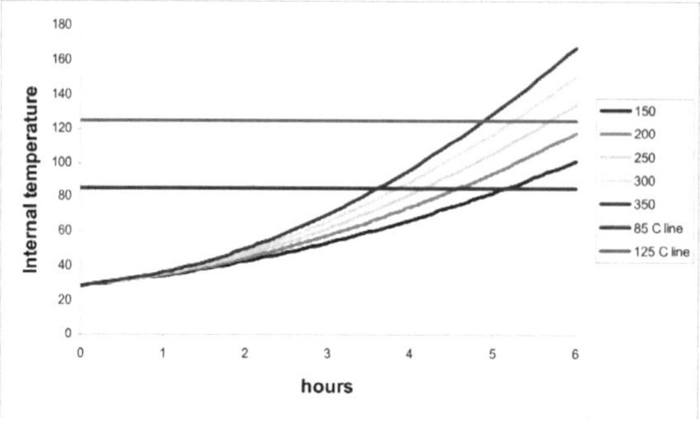

Figure 5. Expected internal temperature vs. time at various external temperatures.

From the plot it can be seen that use of automotive class components will allow for about 5 hours of operation at temperatures up to 300°C. To get to six hours of downhole operation at temperatures greater than 250°C it would appear that high temperature electronics would be called for.

In an attempt to extend the time that the tool can spend in the well, we investigated phase-change cooling. This entails the use of a low melting temperature alloy sealed inside the heat sink. Using this approach, thermal energy would be expended in the phase change of the alloy and delay temperature rise. For comparison, calculations were performed with a low melt alloy, AIM 70, and the results were plotted as seen in figure 6 below.

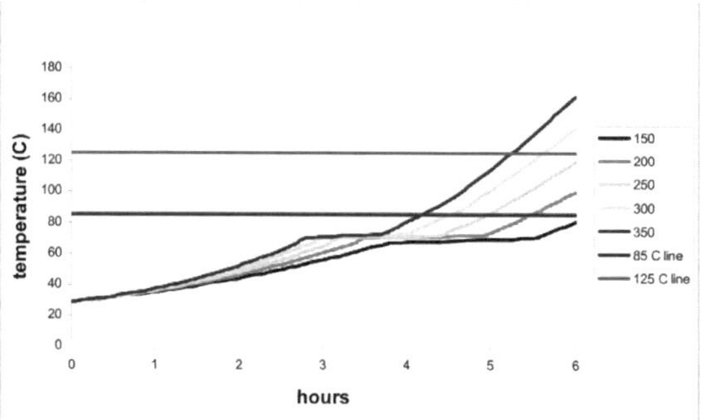

Figure 6. Expected internal temperature vs. time with heat sink that includes phase change material.

From the plot it can be seen that the phase change material would elongate the tool's time in the well by approximately 30-60 minutes depending on the temperature. It is unlikely that we would opt to use this strategy in a prototype tool but it may be part of a future tool. Overall it appears that the main challenges for temperature management include limiting internal heat generation and designing with higher grade automotive components.

Electrical Components

The tool will be controlled by internal electrical components that measure and report tracer signal, temperature and pressure, and control power. These components will be either in the heat-shielded portion of the tool or at the surface. The electronics will be manufactured in a module form allowing flexibility and ease of repair. The added benefit of manufacturing in a module method is that if one module is not performing well it can be swapped out with a better designed module in future iterations of the tool.

Power Supply and Depth Encoder

The power supply will consist of two components. The first is regulation of the power delivered at the surface and the second is the regulation of power at the tool. The powersupply at the surface will convert AC to DC power sent down the wireline. The power supply at the surface will regulate voltage based on the power consumption in the tool and the resistance that is encountered in the line. Figure 7 below shows the relationship of voltage to power consumption with different line resistances encountered. The tool side power supply will be able to give real time information to the surface power supply to determine how much voltage to send.

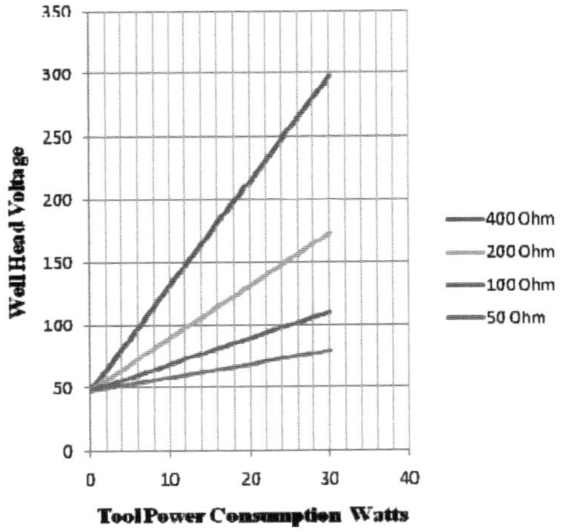

Figure 7. Well head voltage vs. power consumption with different line resistances.

The tool side power supply will involve input protection against over voltage and voltage converters to supply DC power at a variety of voltages to the various components. The power supply will also make use of real time supply optimization to request the right amount of power from moment from the surface supply. This should help to reduce the amount of heat that would be dissipated in the tool. A second strategy that will be used will be turning unused components off when they are not needed. For instance, when the tool is on the way to the zone of interest only intermittent temperature and pressure readings may be desired, the other components can be turned off to reduce internal heat generation.

In addition to supplying power along the wireline, the amount of line paid out will be used to calculate the depth of the tool at any given time. The controls at the surface will contain a depth-encoder device that will be able to accurately calculate the depth of the tool by counting the revolutions of the wireline spool. The depth information will be added to the pressure, temperature, and flow rate data.

Light Source and Detector

The light source is provisionally a high efficiency blue LED that will provide excitation at 465 nm and will operate at 125°C. The LED will be coupled to a collimator which will be connected to the fiber optic source. The LED operation strategy will include modulation to effectively subtract dark current as well as sampling the light to get an accurate indication of intensity out.

The detector will be a photo multiplier tube (PMT). Another detector that was considered was an avalanche photodetectors (APD). The advantages of the APD are low cost, ruggedness and a small electrical footprint. The disadvantages include low sensitivity and inability to operate above 50°C. In contrast the PMT has several orders of magnitude better sensitivity and can operate at temperatures up to 175°C. The disadvantages of PMT's include larger electrical footprint, high voltage (low current) operation and increased cost.

Pressure, Temperature and Inclination

The temperatures within the borehole and internal to the tool will be closely monitored throughout the test. A platinum RTD will be placed through one end of the tools sub-assembly and a second RTD will record temperatures within the tool body. Because some of the components could be destroyed by heat internal temperature will be measured as well. Pressure will be measured using a high temperature piezoelectric pressure transducer using a standard pressure filter port through the tools bottom sub-assembly. An accelerometer will be included in the tool to measure inclination and motion.

Tool Control and Data Transmission

In general much of the tool control will be located within the software being run at the surface monitoring station. Things that will be controlled will be the on/off of individual components, the LED intensity, the PMT gain and the flow rate of the dye pump.

Field Programmable Gate Arrays (FPGA) will be utilized in the communication through the wireline. The FPGA's will allow for flexible signals, in this case data will be packetized and sent with internal validation. Signal prioritization will be controlled by a CAN bus within the tool.

Test and Deployment Plan

The module design of the electronics will afford the ability to test each of the parts independently before attempting to test the tool as an entire unit. The electronics, fiber optics, pump and detector will be tested in autoclave reactors that will simulate geothermal pressures and temperatures. Once the individual components are tested a test fixture will be constructed to test the tool under flowing conditions. A long pipe with inlets at intervals along the length will serve as a test fixture. Water at a specified flow rate can be flowed through the pipe and then water will either be drawn off or added at a known flow rate through the inlets along the length of the pipe. The test fixture will be used to test the tool under flow conditions and serve as a calibration tool for the fluorimeter. Tests in the lab should reveal problems prior to a field test.

Conclusion

The design of a single-conductor wireline fluorimeter-flowmeter for geothermal applications is described. The tool will be useful under both turbulent and laminar flow conditions. Using high temperature electronics, the tool will be able to log well bores in excess of 300°C for periods of greater than 6 hours. It will provide the user with temperature, pressure and volumetric flow rates at each point along the wellbore.

References

Adams, M.C.; Davis J. Kinetics of Fluorescein Decay and its Application as a Geothermal Tracer. *Geothermics* **1991**, *20*, 53-66.

Davarzani, J.; Sloan, M. Simultaneous TPS Logging Systems. *GRC Transactions* **1988**, *12*, 441-448.

de Nevers, N. *Fluid Mechanics for Chemical Engineers*, Third Edition; McGraw Hill: New York, 2005.

U.S. Department of Energy Geothermal Technologies Program (GTP). http://apps1.eere.energy.gov/geothermal/projects/projects.cfm/ProjectID=176 (accessed 27 April 2010).

Henfling, J.A.; Norman, R.A. *Precision Pressure/Temperature Logging Tool*; SAND98-0165; Sandia National Laboratories: Albuquerque, NM, January 1998, 1-21.

Hirtz, P.N.; Kunzman, R.J.; Broaddus, M.L.; Barbitta, J.A. Developments in Tracer Flow Testing for Geothermal Production Engineering. *Geothermics* **2001,** *30*, 727-745.

Lovelock B.G. Steam Flow Measurements Using Alcohol Tracers. *Geothermics* **2001,** *30*, 641-654.

Rose, P.E.; Benoit, W.R.; Adams, M.C. Tracer Testing at Dixie Valley, Nevada Using Pyrene Tetrasulfonate, Amino G, and Fluorescein. *GRC Transactions* **1998**, *22*, 583-587.

Rose, P.E.; Mella, M.; Kasteller, C. A New Tracer for Use in Liquid- Dominated, High Temperature Geothermal Reservoirs. *GRC Transactions* **2003**, *27*, pp 403-406.

Smollen, J.D. *Cased Hole and Production Log Evaluation*; PennWell Books: Tulsa, OK, 1996.

An Integrated Model to Compare Net Electricity Generation for CO_2- and Water-Based Geothermal Systems

Vikas Agarwal[1] and Brian J. Anderson[1]
[1]Department of Chemical Engineering
West Virginia University
Morgantown, WV 26506

Keywords: Wellbore flow, Reservoir simulation, Power cycle, Fluid flow, CO_2, Modeling

ABSTRACT:

Utilization of supercritical CO_2 as a geothermal fluid instead of water has been proposed by Brown[1] and its advantages have been discussed by him and researchers[4,5]. This work assesses the net electricity which could be generated by using supercritical CO_2 as a geothermal working fluid and compares it with water under same reservoir conditions of temperature and pressure. This procedure provides a method of direct comparison of water and CO_2 as geothermal working fluids, in terms of net electricity generation over time.

An integrated three-part model has been developed to determine net electricity generation for CO_2- and water-based geothermal reservoirs. This model consists of a wellbore model, reservoir simulation and surface plant simulation. To determine the bottomhole pressure and temperature of the geothermal fluid (either water or CO_2) in the injection well, a wellbore model was developed using fluid-phase thermodynamic equations of state, fluid dynamics, and heat transfer models. A computer program was developed that solves for the temperature and pressure of the working fluid (either water or CO_2) down the wellbore by simultaneously solving for the fluid thermophysical properties, heat transfer, and frictional losses.

For the reservoir simulation TOUGH2 has been used to model the temperature and pressure characteristics of the working fluid in the reservoir. The EOS1 module of TOUGH2 has been used for the water system and the EOS2 module of the TOUGH2 code has been employed for the CO_2 case.

The surface plant is simulated using CHEMCAD, a chemical process simulator, to determine the net electricity generated. A binary organic Rankine cycle is simulated. The calculated net electricity generated for the optimized water and CO_2 systems are compared over the working time of the reservoir.

---

1. Introduction

There is enough heat in the crust of the earth to satisfy a large part of the world's probable energy needs for a very long time. This heat could be tapped by Enhanced Geothermal Systems (EGS) which aims to extract the heat beneath the earth crust by injecting a fluid through a wellbore, which flows through a hot artificially fractured reservoir and then is forced out from the production well, thus heat is carried to the surface of the earth from beneath the surface. Conventional geothermal power plants are based on extracting the heat from the existing geothermal reservoirs having naturally occurring porosity, permeability which allows the flow of water through them. However, there is vast majority of hot dry rocks that could be found in western U.S.[2] which are assessable through current drilling practice. The heat from these hot dry rocks could be tapped by enhancing the permeability of the rock through hydraulic stimulation and thereafter allowing the fluid to flow through them. One of the most common options is to use water as a heat transfer fluid. However, some of the researchers like Brown[1], Pruess[3] etc. have considered and studied the use of CO_2 as a heat transfer fluid.

Properties that motivate the use of CO_2 as a geothermal fluid emphasized are:

1. The density of cold CO_2 and that of hot CO_2 obtained from production well have large differences, which could provide a significant buoyant force[4] and thus circulation power consumption would be less and more net electricity could be generated.

2. CO_2 is a poor solvent may not dissolve minerals from reservoir thus reducing scaling problems caused by silica dissolution[4]. Conversely water is a powerful solvent, makes difficult to operate EGS reservoir in a stable manner.

3. At the elevated temperature of reservoir, CO_2-mineral reactions may provide an additional advantage for the sequestration of CO_2[5].

This work primarily concerns the comparison of CO_2 and water as a geothermal fluid in terms of net electricity generation over a period of time. The aim of the comparison is to understand under identical set of reservoir conditions which choice is better if one fluid has to be picked over other. A comparison is being made using the same flow rate of both the fluids operating under same surface power cycles and same reservoir conditions compared to work by Pruess et al[3] that compared those under constant delta P.

Any geothermal power plant comprises of three components: the wellbore (injection/production), the geothermal reservoir (from which heat is extracted), and the surface plant (where heat extracted is transformed into electricity for human use). These three components need to be studied separately yet need to be coupled to understand the overall impact on net electricity generation.

In this work, to determine the temperature and pressure profiles of two fluids (water/CO_2) a wellbore model has been developed. TOUGH2, a numerical simulator has been employed for reservoir simulations using the EOS1 module of TOUGH2 for simulating flow of water and EOS2 for CO_2. Finally to determine net electricity generation the chemical process simulation software, Chemcad, has been used for surface plant calculations.

2. Wellbore model

Geothermal fluids either CO_2 or water are being injected into the reservoir through injection wells and the heated fluid are being taken out from the reservoir through production wells. As the fluid travels through the wellbore it either loses enthalpy to or gains enthalpy from the wellbore formation and the rate of loss or gain of enthalpy is a function of time and depth. Therefore geothermal wells should be considered nonisothermal. It is important to know the bottom-hole pressure and temperature in the case of an injection well (to know the enthalpy of the injected fluid to the reservoir) and the wellhead pressure and temperature for the production well. It is the temperature and pressure of the fluid obtained at the wellhead of the production well that determines the amount of electricity that could be generated from the geothermal fluid. Figure 1 shows the basic well configuration used for the modeling. Although the model is developed for any depth and any radius but in this work reservoir is considered to be at a depth of 3 km. Direction of fluid flow in the well could be either upwards or downwards.

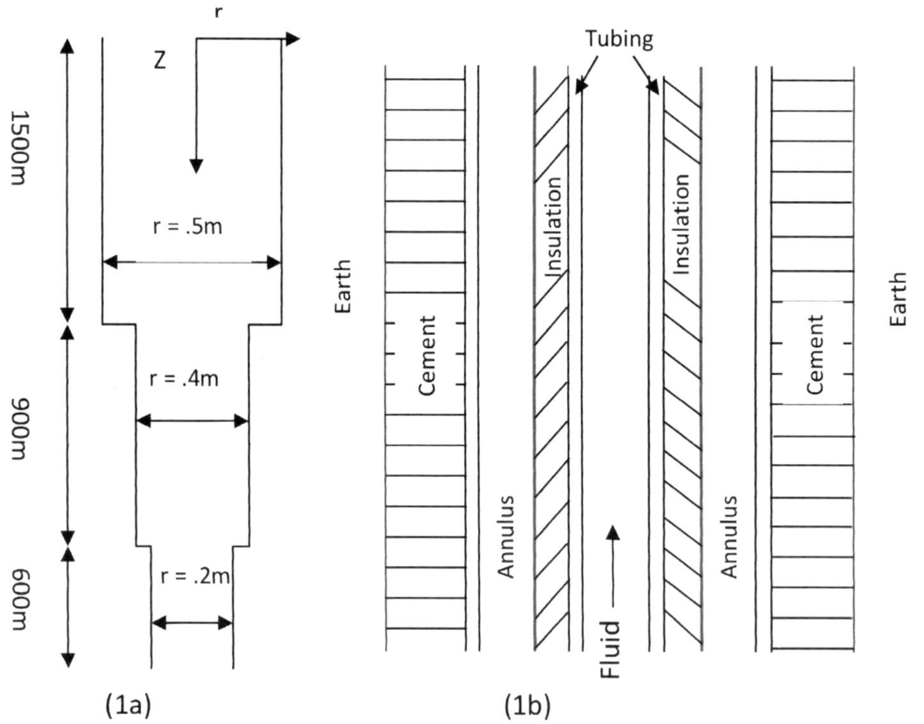

Figure 1: Basic wellbore configuration (1a) Tube diameter variation with depth (1b) Different layers of materials in the radial direction.

Basic flow equations:

We shall now discuss the basic equations used and the methodology in determining the temperature and pressure distribution of the geothermal fluid along the well. The assumptions which are incorporated in the development of this model are:

1) As the diameter of the wellbore is very small when compared to its length, no variation of temperature, pressure and fluid velocity in the radial direction of well is assumed. All properties changes along vertical direction only.
2) For designing flow of CO_2 in the well, the well is taken to be completely dry with respect to water.
3) The geothermal gradient of the earth (which is defined as the rate of change of temperature with depth) is taken to be constant in this work. However, the code could easily be modified to give geothermal gradient variation with depth.
4) The acceleration term of the pressure gradient is neglected. This can be a reasonable assumption, considering that velocity of the fluid doesn't change as fast as the pressure changes with depth. Velocity is a function of density and density is a function of pressure and temperature. As geothermal fluid goes down the wellbore, its pressure increases and hence density tends to increase but due to thermal gradient of earth, temperature also increase which causes the density to decrease and hence the net effect on density and velocity is not that large as in the case of pressure. So, change in velocity is sluggish in comparison to change in pressure. Thus, velocity gradient and hence acceleration term can be neglected in comparison with pressure gradient. It can be justified when the velocity profile is obtained based on this assumption. Based on the similar argument, acceleration term can be dropped when the fluid is ascending the production wellbore.
5) Vertical heat diffusion is neglected as a first effort of model development. It can also be a reasonable assumption because velocity of the fluid is high and convective heat transfer dominates the diffusive heat transfer.

The equation for the pressure drop in a well is given by:

$$\frac{dp}{dZ} + \rho v \frac{dv}{dZ} + \rho g + f \frac{\rho v^2}{2d} = 0 \qquad (1)$$

where,

p = pressure, N/m^2

Z = distance in vertical direction from the wellhead, m

ρ = density of fluid, kg/m^3

v = velocity in z direction, m/s

g = acceleration due to gravity, m/s^2

f = Moody friction factor

d = wellbore diameter, m

All units are taken in SI system.

In Equation (1) the $\rho v \frac{dv}{dZ}$ term represents the acceleration term, ρg represents the gravitational term and $f \frac{\rho v^2}{2d}$ represents the frictional term of the pressure gradient along the well.

Neglecting the acceleration term of the pressure gradient as discussed above, one is left with

$$\frac{dp}{dz} + \rho g + f \frac{\rho v^2}{2d} = 0 \qquad (2)$$

The friction factor, f, can be calculated as[11]

$$f = [-2 \log \{\frac{\epsilon/d}{3.7065} - \frac{5.0452}{Re} \log(\frac{(\epsilon/d)^{1.1098}}{2.8257} + (\frac{7.149}{Re})^{0.8981})\}]^{-2} \qquad (3)$$

Where,

ϵ = Pipe roughness factor

Re = Reynolds number

$$Re = \frac{dv\rho}{\mu} \qquad (4)$$

Temperature calculations:

The variation of the formation temperature along the radial direction as a function of temperature can be given by diffusivity equation:

$$\frac{\partial^2 T_e}{\partial r^2} + \frac{1}{r}\frac{\partial T_e}{\partial r} = \frac{C_e \rho_e}{K_e}\frac{\partial T_e}{\partial t} \qquad (5)$$

where,

T_e = Temperature of earth (°C) at time t

r = Distance from center of well, m

K_e = Thermal conductivity of earth, W/m/ K

C_e = Specific heat of earth, J/kg

ρ_e = Density of earth, kg/m^3

Hasan and Kabir[6] solved the above diffusivity equation using Laplace transform method involving dimensionless variable and concluded the following solution:

$$T_D = \ln\left[e^{-0.2t_D} + (1.5 - 0.3719e^{-t_D})\sqrt{t_D}\right] \tag{6}$$

where,

T_D= Dimensionless temperature which is defined as

$$T_D \equiv -\frac{2\pi K_e}{Q}(T_{wb} - T_{ei}) \tag{7}$$

t_D = Dimensionless time

$$t_D = \frac{\alpha t}{r_{wb}^2} \tag{8}$$

where,

R_{wb} = Outer radius of cement, m

T_{wb}= temperature at R_{wb}, °C

T_{ei} = Initial temperature of the earth as a function of depth, °C

A = Thermal diffusivity, $\alpha = \frac{K_e}{\rho_e C_e}$

T_{ei} is a function of depth and geothermal gradient and can be defined as

$$T_{ei} = T_{eibh} - (L - Z)g_G \tag{9}$$

where,

L = Length of the well, m

T_{eibh} = Bottomhole temperature of the production well, °C

g_G = Geothermal gradient, °C/km

T_{ei} can also be defined on the basis of wellhead temperature in the injection well and geothermal gradient.

To relate the dimensionless temperature (T_D) to the temperature of fluid (T_F) energy balance is carried out as depicted in Figure 2, heat transfers takes place from the earth which is at higher temperature to the cold fluid in the tube of the injection well and vice versa for production well. All resistances to heat transfer across the well, i.e, resistance offered by cement, annulus fluid (air), insulation and tube thickness can be assumed in series.

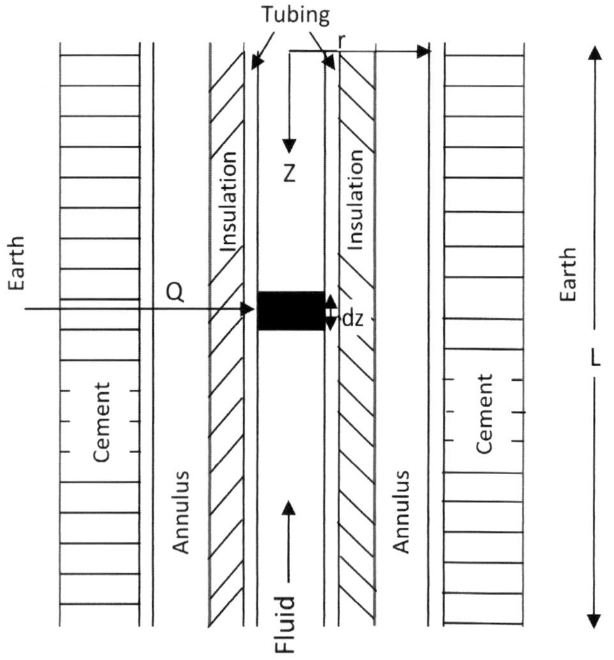

Figure 2: Energy balance on fluid.

Consider an elemental fluid at a depth of z and of control volume Adz, if A is the cross section area. Assuming flow rate of fluid to be *W* kg/s and the specific enthalpy of fluid to be *H* kJ/kg.

Conserving energy on the control volume gives:

$$(WH)_{Z+dz} + Wg(Z+dz) + \left(\frac{wV^2}{2}\right)_{Z+dZ} + Qdz = (WH)_Z + wg(Z) + \left(\frac{wV^2}{2}\right)_Z \quad (10)$$

In Equation (10), the sign of Q changes if the heat is transferred from the fluid to the formation which is the case of injection well.

Rearranging Equation (10) gives

$$\frac{dH}{dZ} + g + V\frac{dV}{dZ} = \pm\frac{Q}{W} \quad (11)$$

Where a negative sign implies production well and positive sign implies injection well.

Substituting the definition of enthalpy as a function of pressure and temperature, Hasan and Kabir[6] obtained the following Equation (12).

$$\frac{dT_f}{dZ} = C_J \frac{dP}{dZ} + \frac{1}{C_P}\left[\mp \frac{Q}{W} - g - V\frac{dV}{dZ}\right] \tag{12}$$

where,

T_f= Temperature of fluid, °C

C_J= Joule- Thompson coefficient, °C/ Pa

C_p = Specific heat of the fluid, J/kg/°C

Heat flow per unit length can be defined as

$$Q = 2\pi r_{to} U_{t0}(T_f - T_{wb}) \tag{13}$$

where,

r_{t0} = Outer radius of tubing, m

U_{t0} = Overall heat transfer coefficient, J/s-m^2-K

For single phase fluid, the Joule Thompson coefficient $C_J$can be approximated as[6]

$$C_J = \frac{1}{\rho C_P} \tag{14}$$

Since the CO_2 is under supercritical conditions throughout the operation of the geothermal energy generation process, it has been assumed that the Joule Thompson coefficient for CO_2 is also given by Equation (14).

Combining Equations 7, 9, 12, 13 & 14 and eliminating T_{wb}, a final equation obtained for production well by Hasan and Kabir[6] used in this model is

$$T_F = T_{eibh} - g_G\left[(L - Z) - (1 - e^{(Z-L)L_r})/L_R\right] \tag{15}$$

Above Equation (15) can be modified for the injection well as follows

$$T_F = T_{eiwh} + g_G\left[(Z) - (1 - e^{(-Z)L_r})/L_R\right] \tag{16}$$

where,

T_{eiwh} = Temperature at the wellhead of the injection well (°C)

L_R is called relaxation parameter and is defined by Ramey[7] as

$$L_R = \frac{2\pi}{C_P W}\left[\frac{r_{to}U_{to}K_e}{K_e+(r_{to}U_{to}T_D)}\right] \quad (17)$$

The correlation for overall heat transfer coefficient is obtained from Hasan and Kabir.[8]

Thermophysical properties of water such as density, viscosity, enthalpy, specific heat etc. are calculated using ASME steam table obtained as an Excel Add-In from Bernhard Spang[9]. For CO_2, Altunin[12] correlations are used, calculated using an Excel Add-In developed by our group. A computer program was developed using Microsoft Visual Basic that solves for the temperature and pressure of the working fluid (either water or CO_2) down the wellbore by simultaneously solving for the fluid thermophysical properties, heat transfer, and frictional losses. Input variables were specified for both water and CO_2 as shown in Table 1 although the code can be used for any flow rate, well radius, geothermal gradient, or injection conditions. A very high geothermal gradient of 80 °C/km is employed in this study to compare H_2O and CO_2 under commercial conditions[2]. The properties of the earth surrounding the well specified in this work are shown in Table 2.

Table 1. Specification of input variables

Well head pressure of the fluid at injection well	8 MPa
Well head temperature of fluid at injection well	35°C
Surface temperature of earth	30°C
Flow rate	15 kg/sec
Geothermal gradient	80°C/km
Well depth	3 km
Inner radius r1	0.5 km
Inner radius r2	0.3 km
Inner radius r3	0.2 km

Table 2. Specification of earth properties

Thermal conductivity of earth	2.42 W/m•K
Diffusivity of earth	10^{-6} m^2/sec
Thermal conductivity of cement	1.50 W/m•K
Annulus fluid	Air

Results obtained after running the code for various times are shown in Figure 3a, 3b and 3c.

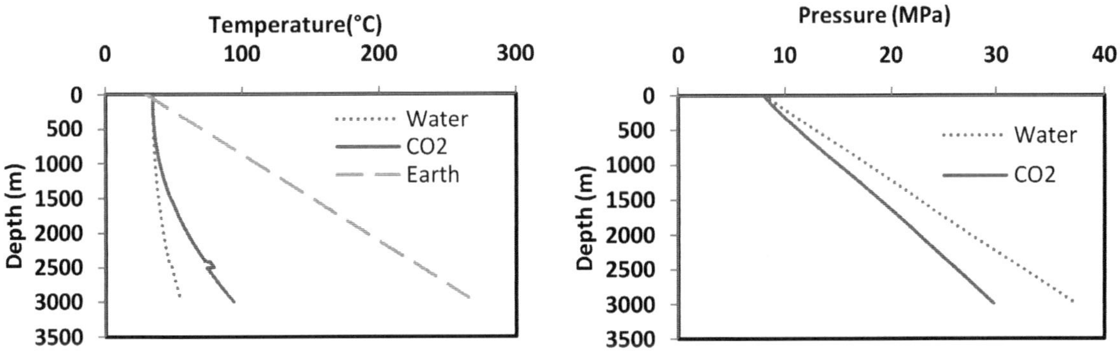

Figure 3a: Temperature and pressure profiles of CO_2 and water in the injection well after 0.001 years.

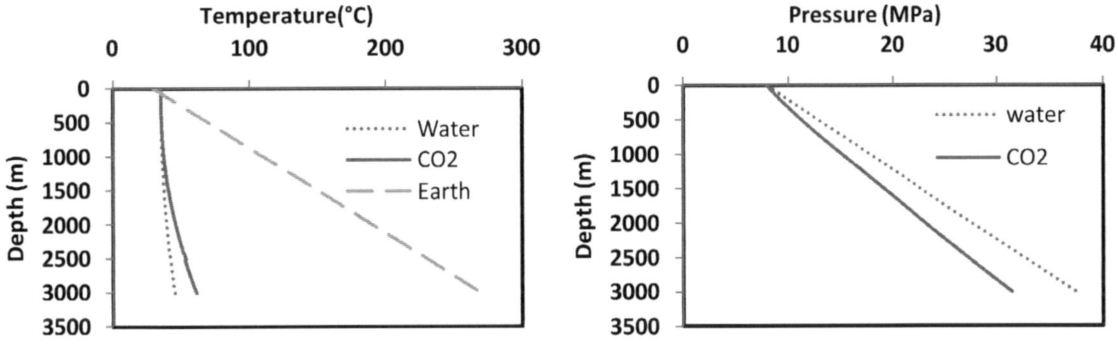

Figure 3b: Temperature and pressure profiles of CO_2 and water in the injection well after 10 years.

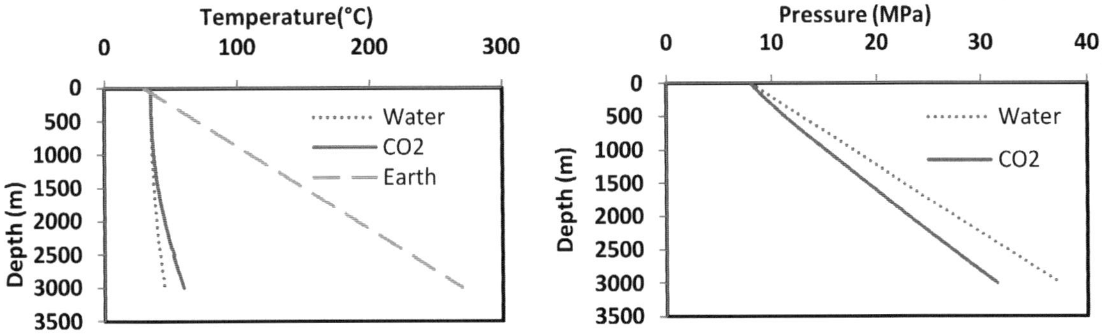

Figure 3c: Temperature and pressure profiles of CO_2 and water in the injection well after 30 years.

As shown in the Figure 3 and also expected theoretically, CO_2 gains more temperature from the surrounding formation than water due to the virtue of its low specific heat capacity. On the same basis, it loses more temperature than water in the production well. As time progress temperature of earth adjacent to the injection well (see Figure 1) starts to decrease and thus the rate of heat gained by the fluids (both water and CO_2) from the earth starts to decrease.

The increase in pressure for water is more than that of CO_2 because of higher density of water than supercritical CO_2. The net gain in pressure decreases over time for both the fluids. The bottom-hole temperature and pressure of the injection well obtained from the wellbore model is subsequently used in the reservoir simulation.

3. Reservoir simulation

A five spot pattern is considered for reservoir simulation as shown in Figure 4. Due to the high level of symmetry only 1/8 of the whole reservoir needs to be modeled. As mentioned before, a flow rate of 15 kg/sec is being injected on the basis of 1/8 of the reservoir, thus on the full well basis a flow rate of 120 kg/sec is being injected. All results shown in this work are on the basis of 15 kg/sec.

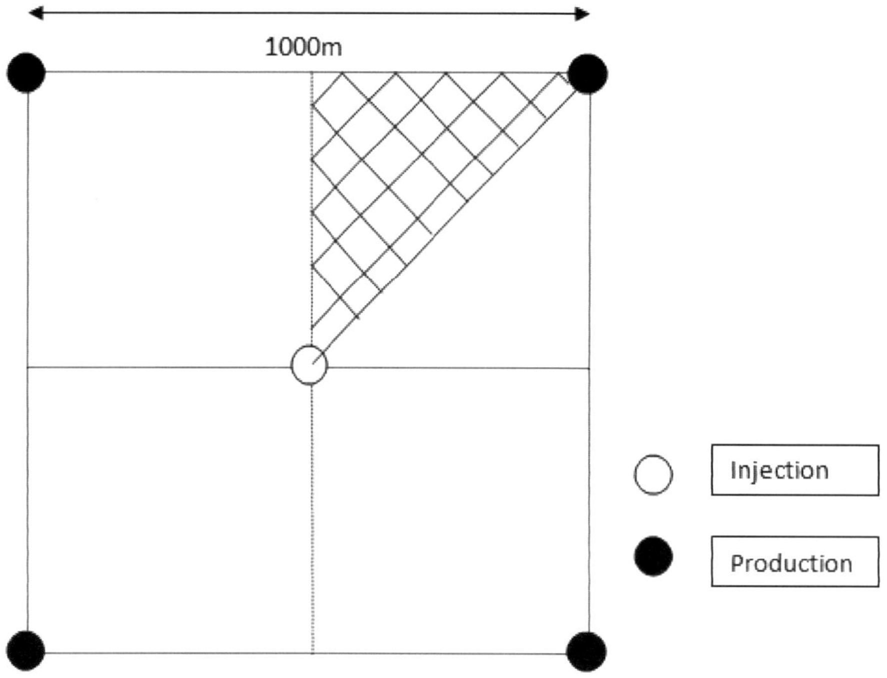

Figure 4: Five spot pattern reservoir

Parameters used for reservoir simulation using TOUGH2-MP are shown in Table 3. These are based on the parameters used by Karsten Pruess[3].

Table 3. Parameters of the reservoir	
Thickness	305 m
Grid block dimensions	70.71 m ×70.71 m
Permeability	60×10^{-15} m^2
Rock Density	2650 kg/m^3
Rock specific heat	1000 J/kg/°C
Rock thermal conductivity	2.1 W/m/°C
Initial reservoir temperature	270°C
Initial reservoir pressure	32 MPa

Results obtained from reservoir simulation are shown in Figure 5. It is assumed that the temperature of the fluid coming out of the production block is same as the production block.

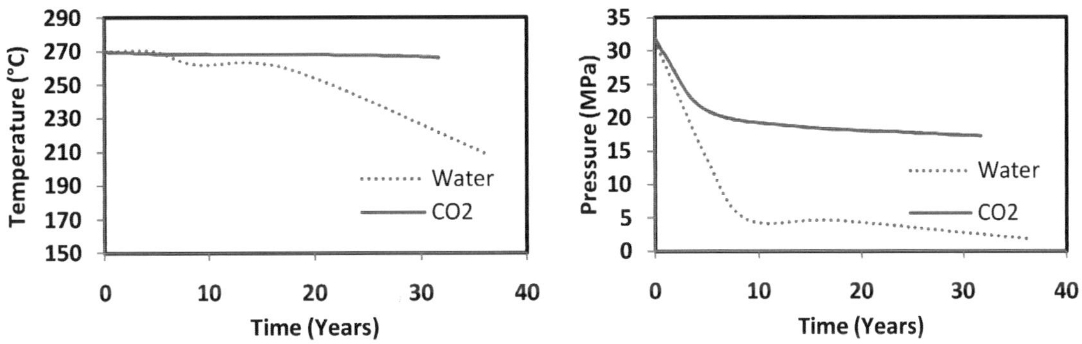

Figure 5: Temperature (left) and pressure (right) of the production block with time.

Results showed that more pressure and temperature drop is observed in the reservoir using water as a flowing fluid than CO_2 having exactly same flow rate. Thus, it can be concluded that there is more heat sweep of the reservoir having water as a heat transmission fluid.

It is required to know the temperature and pressure of water and CO_2 at the top of the production well. For this wellbore model is being used. The bottomhole temperature and pressure obtained from reservoir simulation results at the production block served as the input parameter for the wellbore model. Results obtained from wellbore model are shown in Figure 6.

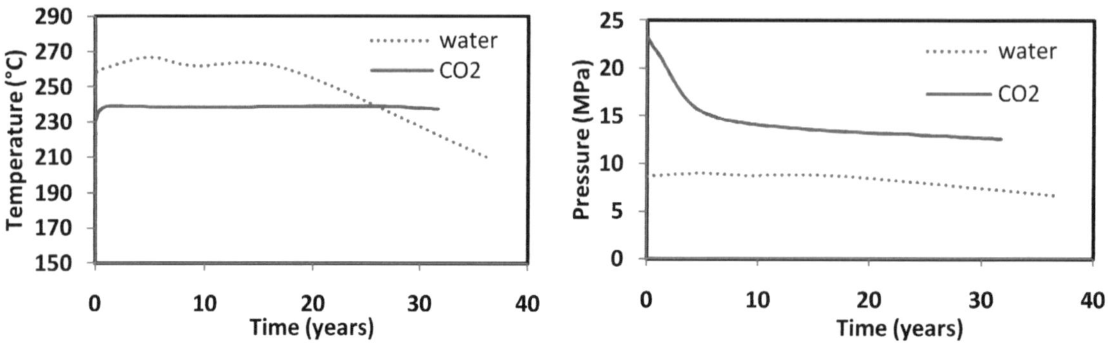

Figure 6: Temperature (left) and pressure (right) at wellhead of the production well with time.

It is important to note that a pump has been placed for water in order to avoid two phase flow in the well as the surface plant is designed for only hot pressurized water and not for wet steam. The role of pump is to increase the pressure of the water from the production block back to 32 MPa. However, amount of work consumed by the pump has been accounted in the net electricity generation calculations.

4. Surface plant

The temperature and pressure of the geothermal fluid over time at the wellhead of the production well determines the amount of net electricity generation. A surface plant is designed based on the Rankine cycle to determine electricity production. In this cycle, an organic fluid such as isobutene, iso-pentane, R-32 (refrigerant) etc. takes heat from the geothermal fluid (water or CO_2) and is then get vaporized because of low boiling point. This high-pressure, vaporized organic fluid is then passed through the turbine/generation unit and work (electricity) is obtained. After coming out of the turbine it is then condensed in the condenser and then again pumped back to the heat exchanger (vaporizer) where it gets vaporized and thus the cycle is completed. An extra heat exchanger has been used in the present work in order to extract more heat from the geothermal fluid coming out of the first heat exchanger. The amount of work obtained through this cycle will also depend upon the choice of organic fluid. As mentioned by the Augustine et al.[10], considering the temperature of the geothermal fluid at the wellhead of the production well, the maximum efficiency would be obtained using iso-pentane as the secondary fluid. Hence, iso-pentane is the secondary fluid used in the Rankine cycle. Figure 7 shows the schematic diagram of the surface power plant being simulated.

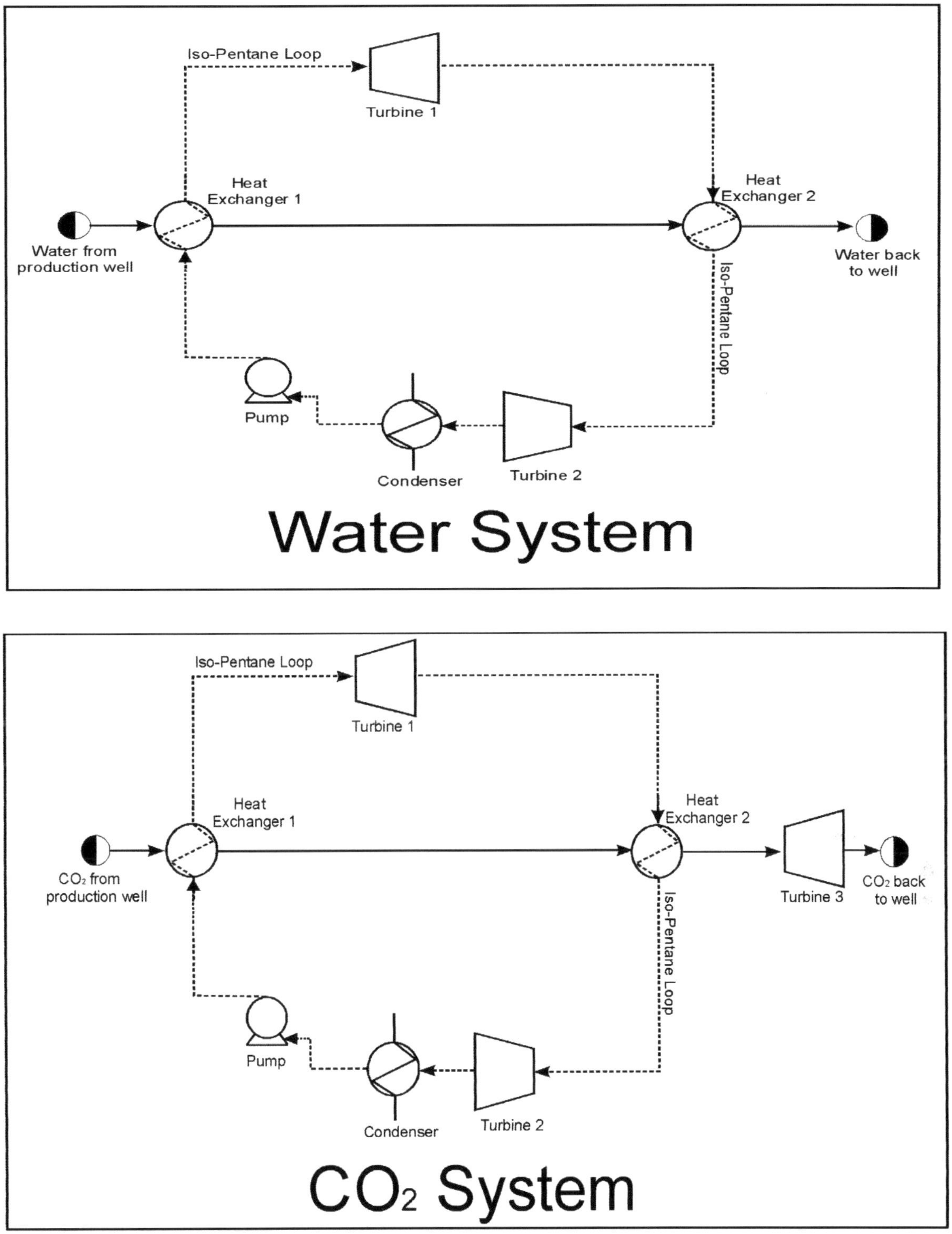

Figure 7: Schematic diagram of power plant for water (Top) and CO_2 (Bottom) as the heat transmission fluid.

It is worth noting that an extra turbine could be employed for CO_2 as it coming out of the production well at a higher pressure and would add more to the net electricity generation. The cycle is being operated in such a way that the heat transmission fluid either water/ CO_2 coming out of the cycle nearly equals the injection conditions at the injection well, hence overall cycle is completed. Chemcad 6 has been used to calculate net electricity obtained through the cycle. The overall results are shown in Figure 8. Theoretically, water produces more electricity than CO_2, but the rate of the decline in net power generation is found to be much more rapid for water than CO_2 as shown in Figure 8.

Although CO_2 has lower pressure drop in the geothermal well and reservoir, but due to its lower heat carrying capacity, less heat is being extracted from the reservoir and thus less power is being produced.

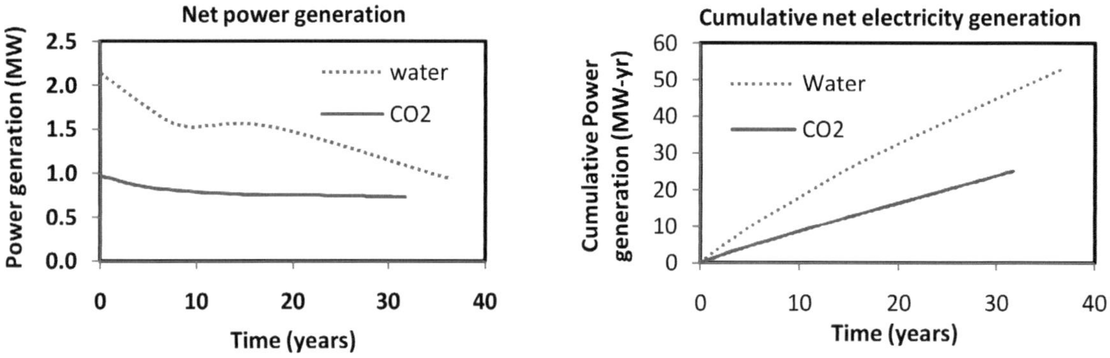

Figure 8: Net power (MW) and net cumulative electricity (MW-yr) could be generated over time.

5. Conclusion

Based on theoretical calculations in this work and other specifications assumed, water is found to produce nearly 70-80% more net power in the initial 10 years of power generation than CO_2 keeping all other parameters exactly same. However, power generation employing water reduces considerably over time giving a disadvantage over CO_2 which seems to produce nearly consistent power generation. After 30 year of plant operation water produces 80% more cumulative electricity than CO_2.

A important conclusion about CO_2 could be drawn is that it may not require external power circulation as found in this work and the flow of CO_2 through the complete geothermal cycle is possible just on the basis of density difference of cold and hot supercritical CO_2. So, it could be used as a geothermal fluid in the areas having higher geothermal gradient but facing water scarcity problems.

Loss of water in the reservoir would cost money but loss of CO_2 could be a mean of CO_2 sequestration and hence could earn carbon credits. More frequent fouling in heat exchangers and other units of power plant could be seen employing water because of high minerals dissolved in it, however with CO_2 it might not be the case.

Karsten Pruess[3] in his work has shown that approximately 3.7 times flow rate of CO_2 could be achieved over water keeping the pressure drop of the reservoir to be same and further concluded that CO_2 has more heat extraction rate over water under constant bottom-hole conditions of injection and production well. However, in this study, flow rate of both the fluids are kept same and the more pressure of CO_2 obtained over water (at the production block of the reservoir) is being harnessed at the surface power plant by providing an extra turbine. Further studies are being conducted on the effect of CO_2 flow rates on net electricity production.

References:

1. Brown, D. A Hot Dry Rock geothermal energy concept utilizing supercritical CO2 instead of water. In: Proceedings of the Twenty-Fifth workshop on Geothermal Reservoir Engineering, Stanford University, Pages 233-238, 2000
2. U.S. Department of Energy, Geothermal Technologies Program-Energy Efficiency and Renewable Energy.
3. Pruess, K., Enhanced geothermal systems(EGS) using CO_2 as working fluid- A novel approach for generating energy with simultaneous sequestration of carbon, Geothermics, Volume 35, Pages 351-367, 2006.
4. Fouillac, C., Sanjuan, B.,Gentier, S., Czernichowski-Lauriol, I., Could sequestration of CO2 be combined with the development of Enhanced Geothermal Systems? Third Annual Conference on carbon capture and Sequestration, Alexandria, VA, 2004.
5. Metz, B., Davidson, O., De Coninck, H., Loos, M., Meyer, L., IPCC Special Report on Carbon Dioxide Capture and Storage, Intergovernmental Panel on Climate Change; Geneva (Switzerland); Working Group Iii, Cambridge University Press, 2005.
6. Hasan, A.R. and Kabir, C.S. – Fluid Flow and Heat Transfer in wellbores, Chapter 5, Pages 65-69.
7. Ramey, H.J., Wellbore Heat Transmission, Trans., AIME, 225, 1962.
8. Hasan, A.R. and Kabir, C.S. – Fluid Flow and Heat Transfer in wellbores, Appendix-A, Pages 151-153

9. Bernhard Spang , Water97-v13.xla-Excel Excel Add-In for properties of Water and Steam in SI-Units, Hamburg, Germany, February 2002 http://www.cheresources.com/iapwsif97.shtml
10. Augustine, C., Filed, R., DiPippo, R., Gigliucci, G., Fastelli, I., Tester, J., Modeling and Analysis of Sub- and Supercritical Binary Rankine Cycles for Low- to Mid- temperature Geothermal Resources, GRC Transactions, Vol.33, 2009.
11. Chen, N.H., An Explicit Equation for Friction Factor in Pipe, Ind. Eng. Chem. Fund., 18, 296, 1979.
12. Altunin, V.V., Thermophysical Properties of Carbon dioxide, publishing house of standards, Moscow, 1975.

Assessment of Effective Heat Transfer Area with Sorbing Tracers

Kevin Leecaster[1]
Pete Rose[1]
Susan Petty[2]
[1]Energy & Geosciences Institute/Univ. of Utah
423 Wakara Way, Suite 300
Salt Lake City, Utah 84108
801 585 0793
kleecaster@egi.utah.edu

[2]AltaRock Energy, Inc.
7900 E. Greenlake Way North Suite 202
Seattle, WA 98103

Assessing the surface area from which the working fluid draws heat in geothermal energy extraction systems could help with the management of existing resources and with the design and construction of engineered geothermal systems in hot dry rocks. In otherwise identical aquifers hosting coupled injection & production wells, a difference in the amount of surface area available to heat the injected fluid will influence the amount of energy that can be extracted without a loss in fluid enthalpy at the production well occurring. Thermal drawdown can occur when the heat extraction rate through fluid convection exceeds the conduction rate from the host rocks to the fracture surfaces where most heat transfer occurs. Predicting the host rock thermal response to the induced fluid convection requires an estimation of, among other things, the effective surface area available to transfer heat to the fluid. Operating geothermal reservoirs often have flow paths plugged by mineral precipitation near the injection wells. This decreases the available heat transfer area and could lead to premature thermal drawdown if not identified and addressed. Creation of new fractures within hot dry rocks will open up unconventional geothermal reservoirs. A tool to assess the exposed surface area within these new reservoirs would help to support this advancing technology. Since adsorption of solutes is a function (in part) of how much surface area/fluid interaction occurs in a system, sequential tracer breakthrough curves through the same system with only a change in its surface area is a potential surface area measurement tool.

We are working on developing thermally stable tracers that will help geothermal energy producers to assess the surface area available between injection and production wells. Research results are presented that show how the use of tracers that interact (i.e. sorb) with the surface of the geomedia can help to refine the estimation of the available effective heat transfer area. A series of high temperature miscible displacement experiments through media with different size fractions of Ottawa sand demonstrates the relationship between the retardation of a reactive tracer relative to a conservative tracer and the amount of surface area that the fluid encounters. Thermal characteristics of the tracer and temperature effects on its retardation through the media are presented, as well as a discussion on analysis, detectability, and availability of potential reactive tracer candidates for field applications.

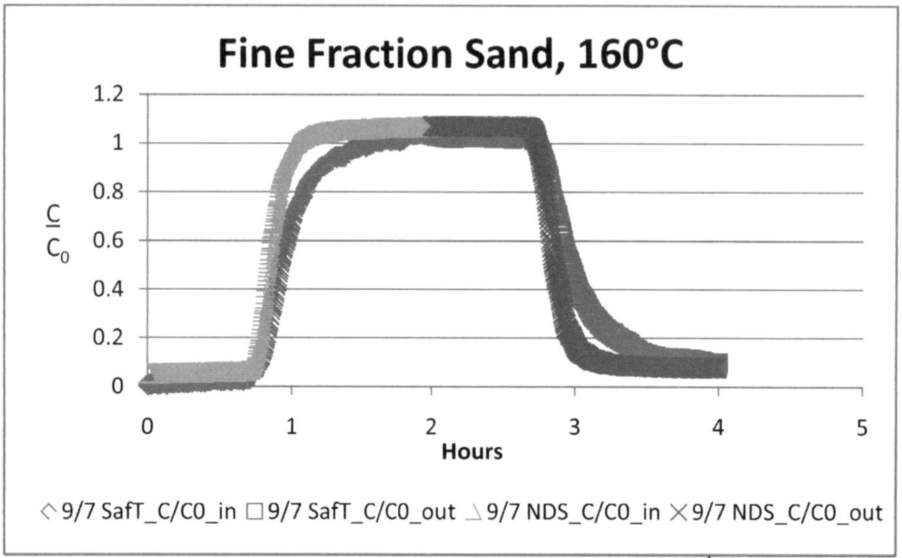

Figure 1 Breakthrough curves from step input of 1.4 mg L^{-1} safranin T and 100 mg L^{-1} 1,5 naphthalenedisulfonate in 200 mg L^{-1} NaCl, 175 mg L^{-1} $CaCl_2$, and 10 mg L^{-1} tetraethylorthosilicate flowing at 2 mL min^{-1} through Ottawa sand fraction of 180 to 425 μm at 160°C.

Two food colorings, yellow #5 & blue #1, and three safranines were examined for thermal stability. Our tests indicate that the blue #1 is more themally stable than the yellow dye, but only up to a working temperature of around 150°C. Of the tested molecules trimethyl safranin proved the most thermally stable (table 1). Our results indicate that it could be a viable option in reservoirs hotter than 220°C. The miscible displacement experiments show that safranin is retarded relative to a conservative tracer at temperatures up to 160°C (figure 1). The safranin T in that experiment had a retardation factor of 1.12 with a standard deviation of 0.008 from two experiments. Additionally, results shown in figure 2 demonstrate a difference in the retardation factor derived from breakthrough curves from media of the same mineralogy but different particle sizes over three different temperatures. The retardation of the reactive tracer decreases at higher temperatures, but distinctly shows greater retardation with smaller particle sizes (i.e. greater surface area).

Table 1 Thermal Stabilities

Tracer	Temperature (°C)	Hours at Temperature	Degradation Rate (percent loss/day)
Yellow #5	150	41	57%(±0.1%)
Blue #1	150	41	11%(±0.9%)
Phenosafranin	140	26	21%(±5.6%)
Safranin O	150	72	14%(±1.8%)
Safranin T	220	136	6% (±2.3%)

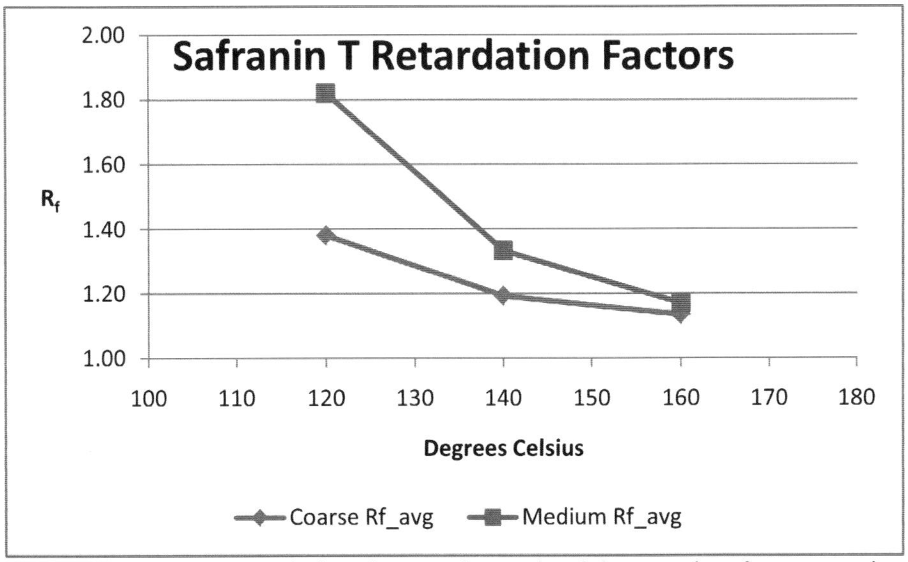

Figure 2 Average retardation factors determined from ratio of conservative and reactive tracers' median retention times. Coarse fraction sand is between 600 and 500 μm and medium fraction is between 500 and 425 μm.

THE CHEMICAL AND THERMAL STABILITY OF PROPPANTS UNDER GEOTHERMAL CONDITIONS

Kristie McLin[1], Daniel Brinton[1,2], Prashanth Mandalaparty[2], Clay Jones[1], and Joseph Moore[1]

1. Energy & Geoscience Institute, University of Utah, Salt Lake City, UT 84108
2. Dept. of Chemical Engineering, University of Utah, Salt Lake City, UT 84108

Keywords: EGS, fracture, proppant, reservoir engineering

Abstract

Enhanced geothermal system (EGS) reservoir fracture creation and management may require the use of proppants to maintain fracture conductivity. Hydraulic fracturing accompanied by the addition of proppant is a common procedure in the oil and gas industry. In contrast, the use of proppants by the geothermal industry has been limited. A variety of proppants are available, and the most commonly used consist of silica sand, ceramic, resin coated sands, and bauxite. These solids remain in newly formed fractures to keep them open. In geothermal systems, proppant will need to withstand high temperatures, acidified fluids, acid treatments, and cleanouts while maintaining the porosity and permeability of the fracture. However, proppant particles may act as nucleation sites and promote precipitation, or they may dissolve, consequently affecting fracture performance. Thus, the performance of proppants in geothermal reservoirs must be understood to avoid costly mistakes in the creation and maintenance of EGS reservoirs.

A series of static experiments have been conducted at 200°C using 30/60 Sinterball bauxite proppant provided by Sintex Minerals. The proppant consists of corundum with minor quartz, hematite and chlorite. De-ionized water has been used for the fluid phase. The experiments were run for periods of one to two months.

X-ray diffraction, scanning electron microscope and QEMScan analyses were performed on the proppant before and after testing to evaluate textural, mineralogical and compositional changes that occurred as a result of water rock interactions. Only minor changes were observed. These changes include the presence of surface pitting, resulting from dissolution of impurities in the proppant grains and deposition of minor amorphous silicates and aluminum oxides.

Introduction

In the oil and gas industry, hydraulic fracturing is often accompanied by the addition of proppants. Proppants are sized particles that are added to fracturing fluids and remain in the fracture to physically hold the newly created fracture open while maintaining conductivity within the fracture. A variety of proppants are available, and the most commonly used proppants consist of silica sand, ceramic and resin coated sands, and bauxite. In enhanced geothermal systems (EGS), proppant will need to withstand high temperatures, acidified fluids, acid treatments, and cleanouts while maintaining the porosity and permeability of the fracture. However, proppant particles may act as nucleation sites and promote precipitation, or they may dissolve, consequently affecting fracture performance. Thus, the performance of proppants in geothermal reservoirs must be understood to avoid costly mistakes in the creation and maintenance of EGS reservoirs. The goal of this investigation is to develop improved methods for maintaining permeable fracture volumes in EGS reservoirs by experimentally evaluating the performance of proppants (natural and manufactured) under conditions found within geothermal environments and determining the effects of mineral deposition and dissolution on propped fractures.

Previous Studies

A study in the Northeast German Basin used high-strength ceramic proppant in a well at 4080-4190m depth and 140°C (Legarth et al, 2003). Propped fractures were created, and inflow performance of well was significantly enhanced. Even so, post-frac productivity was not as good as had been predicted. The possible reasons given for this discrepancy are chemical and mechanical processes during shut-in or undersized and poorly connected fractures.

Previous studies from the oil and gas industry have reported that sintered bauxite proppant is resilient under high stress environments (Cooke, 1977). When formation stresses are high, these proppants deform, rather than crush. This deformation allows for the increase of contact area between proppant grains or between proppant and the wall rock. Sand has been shown to crush, decreasing porosity and permeability in the fracture (Cooke, 1977).

Experimental Setup

A variety of proppants are currently available on the market. These include silica sands, ceramic proppants and bauxite proppants. Our current series of tests have utilized Sintex Minerals and Services, Inc. Sinterball Bauxite proppant. SinterBall Bauxite 30/60 is a high strength bauxite proppant used for high stress, temperature, and flow rate hydraulic fractures. The high alumina content of this bauxite is what provides the strength, chemical resistance and thermal stability it needs to ensure optimum production. Mechanical properties are shown in Table 1a-c, and the relationship between closure stress and permeability is shown in Figure 1.

Table 1a-c. Mechanical properties of the SinterBall Bauxite 30/60 proppant as provided by Sintex Minerals and Services, Inc.

a.

Technical Data	
Specific Graviy	3.63
Bulk Density	130 lb/ft^3
Crush @ 12,500 psi	2.1%
Absolute Volume	.033 gal/lb
Mean Particle Diameter	.501 mm

b.

Sieve	Retained
30	3%
40	77%
50	18%
60	2%

c.

Closure Stress (psi)	Permeability (darcies)	Conductivity (md-ft)	Beta Factor (atm-sec^2/kg)
6,000	169	2,342	0.376
8,000	137	1,854	0.422
10,000	120	1,580	0.470
12,000	103	1,321	0.581
14,000	86	1,073	0.741

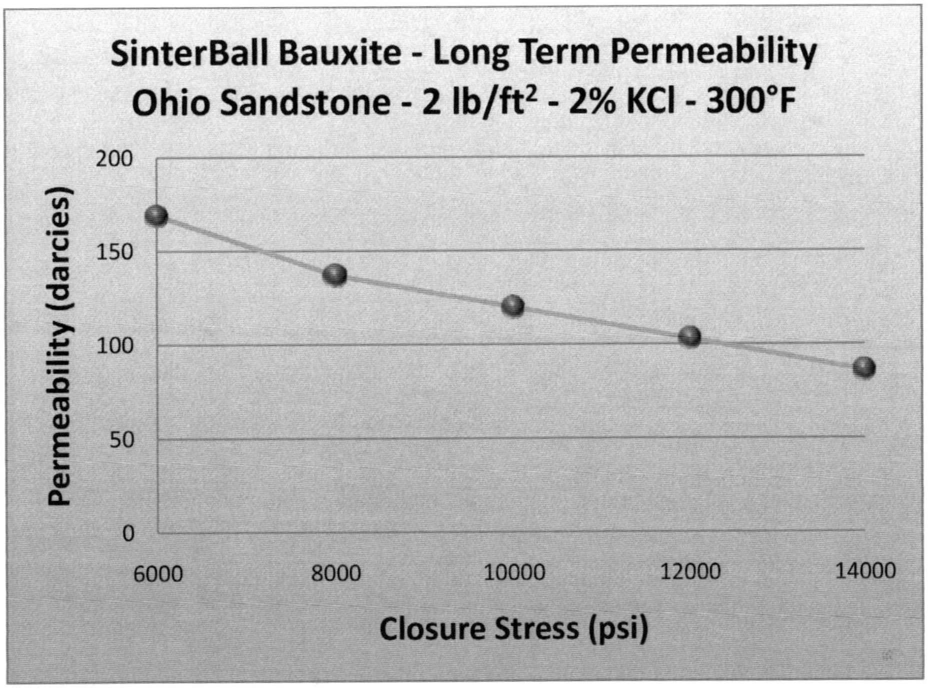

Figure 1. Relationship of closure stress to permeability of the SinterBall Bauxite 30/60 proppant as provided by Sintex Minerals and Services, Inc.

The experimental apparatus is shown in Figure 2. It consists of a stainless steel reactor made of 316-grade stainless steel, which was rated for 4000 psi at 600°C. The reactor is heated by means of electrical heat tape. The temperature is controlled via a k-type thermocouple using a temperature controller and a SPECVIEW interface. The reactor was designed with removable top and bottom caps which allow for solids to be easily added and removed. The experimental apparatus is modified from the previous experimental setup described in Seyfried et al. (1987), which was adopted by Kaszuba et al. (2003, 2005) and Rosenbauer et al. (2005). Experiments were conducted at 200°C. The first two experiments used de-ionized water as the liquid phase.

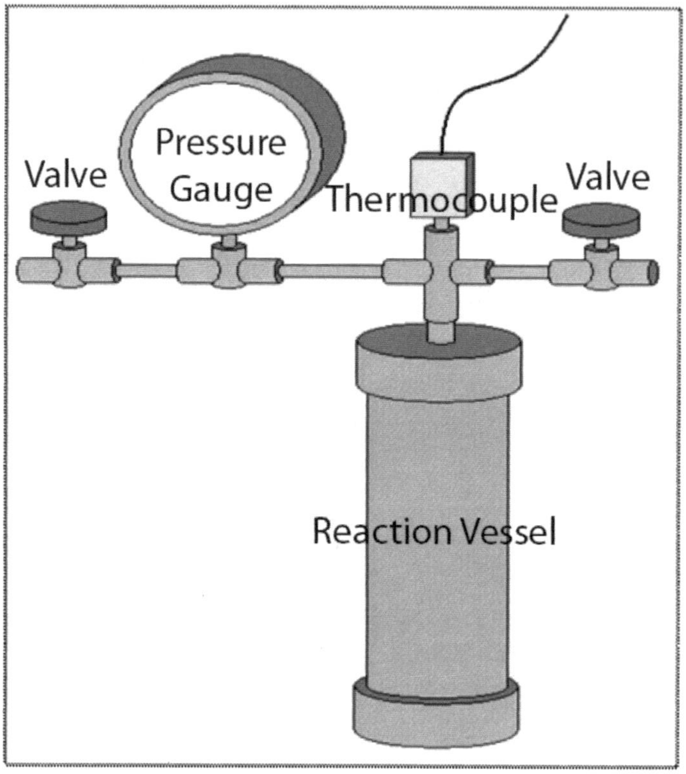

Figure 2. Reactor vessel setup.

Batch experiments on Sintex 30/60 proppant were conducted using de-ionized water at 200°C. Experiments were conducted in identical reactors. The initial samples were prepared by weighing an appropriate amount of proppant and placing it into the reactor. DI water was slowly added to the reactor to allow the water to fill the pore spaces until the water level was just below the top of the reactor. The reactors were then sealed and held at a constant temperature of 200°C for one to two months.

After the reaction period, the samples were retrieved from the reactor. The proppant and other newly precipitated phases were retrieved by removing the end caps and removing the sample, being careful to minimize disturbance to mineral precipitates and crystals. Reactor liquids were separated from the solids by vacuum filtration with Whatman 40 filter paper. The solids were air dried and analyzed using SEM/EDS, XRD and QEMScan. Reactor liquids were acidified with sulfuric acid to prevent any further precipitation and were analyzed using ICPMS to determine the concentrations of ions. Preliminary SEM and XRD analyses were conducted on the dried reactor solids to identify the precipitated phases and dissolution patterns in the mineral matrix after the experiments. The SEM analysis was done at different resolutions to examine the textural features and chemical composition of the proppant prior to and after reaction. In these experiments, the reactor systems were allowed to cool over night before liquids and solids were removed. This was done to avoid the flashing of reactor liquids resulting in precipitation of dissolved solids. However, the extended cooling time was likely sufficient for significant precipitation to occur even in the absence of flashing.

Results

A gray clay mineral was observed in reaction vessels at the end of the first two experiments. Aggregates of this mineral were found near the base of the first reactor. In contrast, this mineral was dispersed throughout the proppant in the second reactor. This mineral was not observed in the un-reacted proppant. In addition to the differences in the distribution of the clay mineral, the proppant at the base of the second reactor had formed a dense hard mass that was cemented to the walls of the vessel. Efforts to dislodge the proppant were largely unsuccessful. X-ray diffraction (XRD), scanning electron microscope (SEM), and QEMScan analyses were performed on the proppant before and after reaction to evaluate textural, mineralogical and compositional changes that occurred as a result of water rock interactions.

Scanning Electron Microscope Analyses

An FEI SEM was used to image the un-reacted and reacted proppant. The proppant was left uncoated to increase the accuracy in chemical analyses using an energy dispersive X-ray unit (EDX). The un-reacted proppant is shown in Figures 3a-c. XRD analysis shows that the proppant is mostly corundum with minor quartz and hematite. Figure 3c shows a close up of the un-reacted proppant and the corundum plates. There are occasional pits in the un-reacted proppant, and these pits are more iron rich than the rest of the proppant surface.

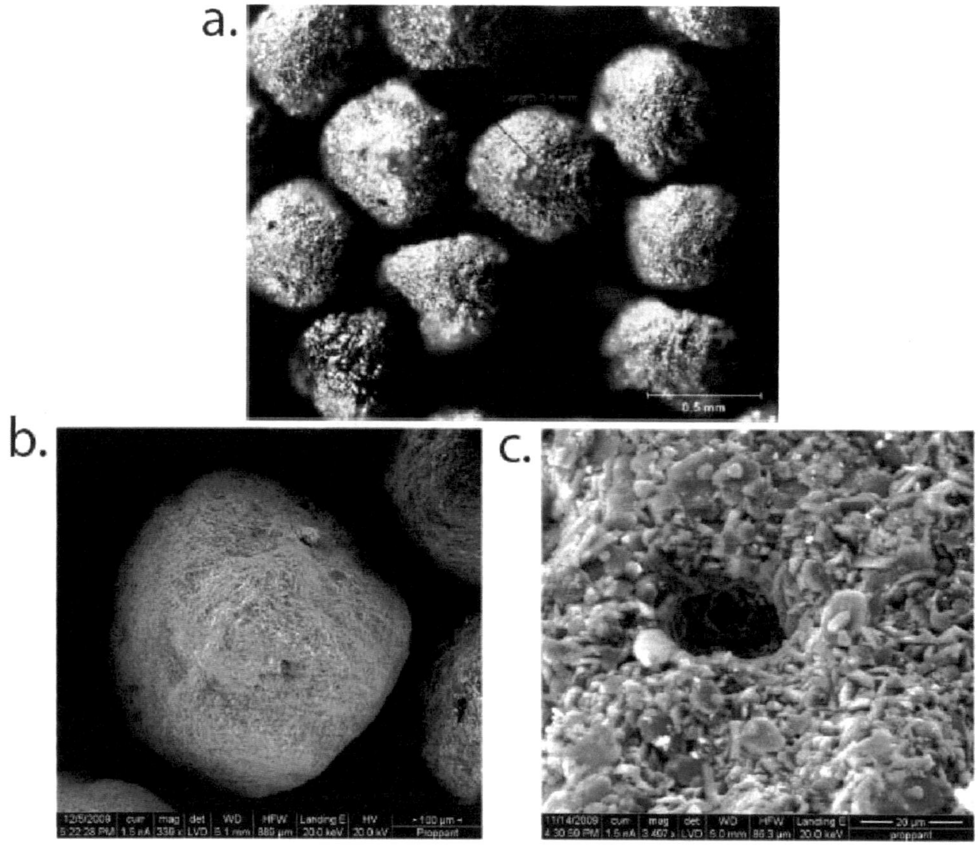

Figure 3a-c. (a). Magnified untested proppant. (b). SEM image of untested proppant, showing minor surface pitting. (c). SEM image of untested proppant showing corundum plates and close up of pitted surface.

The proppant that was reacted in experiment 1 is shown in Figures 4a, b. It can be seen that the surface texture has changed and that dissolution of the corundum plates has occurred (Fig. 4b).

Clay minerals (Fig. 5) were found in the bottom of the reaction vessel. X-ray diffraction analyses indicate these clays are Mg and Fe rich chlorites.

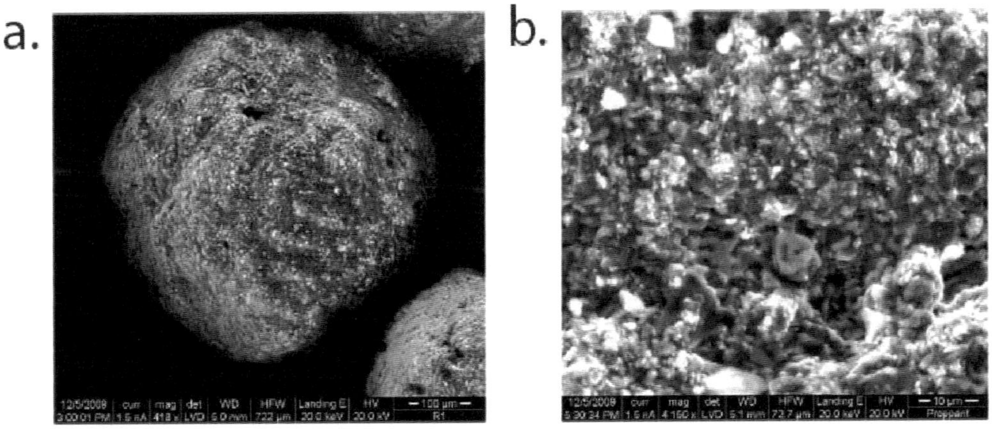

Figure 4a, b. (a). SEM image of proppant reacted in experiment 1. (b). Close up SEM image of proppant reacted in experiment 1 showing dissolution textures on the corundum plates.

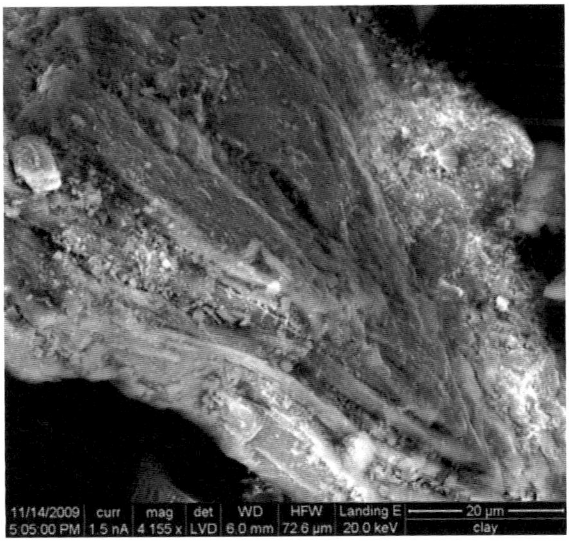

Figure 5. SEM image of clay found in the bottom of the reaction vessel from experiment 1.

The proppant from experiment 2 shows similar dissolution, but it has also been coated with a variety of other mineral phases. Figure 6a shows the texture of the "gibbsite". Gibbsite is a hydrate of alumina and is common in bauxites with diaspore and boehmite. We use the term "gibbsite" to represent these phases, as they have not been positively identified by X-ray diffractometry. Figures 6b,c show a silicate coating on proppant from the reacted sample. The texture of the silica phase suggests that it amorphous.

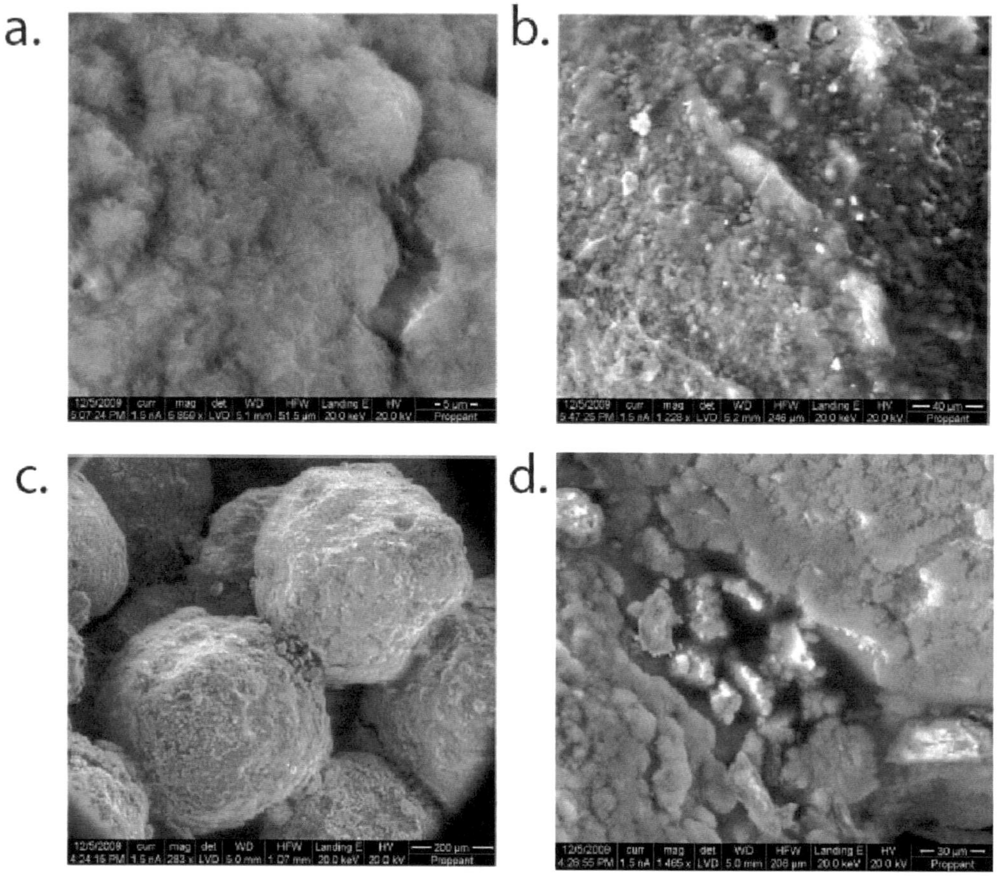

Figure 6a-d. SEM images of proppant from experiment 2. (a). Aluminum oxide coating on proppant. (b). Layered coatings of aluminum oxide and amorphous silicate on proppant. (c). Amorphous silicate cementing two proppant grains together. (d). Close up image of the amorphous silicate cement from (c).

X-Ray Diffraction Analyses

The XRD analyses provide additional detail on the mineralogy of the samples. Figure 7 shows the XRD scans and mineralogy of the un-reacted and reacted proppant samples from experiment 1 and experiment 2 (top and bottom of reaction vessel). There is no significant change observed in the bulk mineralogy of the samples. It is possible that other minerals are present in the sample with concentrations of less than a few percent, but are not visible because: (1) they do not have a strong primary peak making them indistinguishable from the background; (2) their peaks overlap with another mineral of higher concentration; or, (3) they are iron rich and therefore are prone to absorb X-rays. Figure 7a shows the glycolated clay sized fraction of the un-reacted and reacted samples from experiment 1 and experiment 2 (top and bottom of reaction vessel). There are no prominent peaks at low angles indicting well formed sheet silicates are of low abundance. Chlorite, illite, and aluminum hydroxide (bauxite) minerals occur in trace amounts. The broad hump at low angles suggests that a poorly formed "clay-like" mineral may be present. Several subtle differences can be observed in the glycolated clay sized X-ray diffraction patterns. A secondary quartz peak at about 21 2 theta is not present in the 'Experiment 2 Bottom' sample indicating a lower abundance of quartz; and primary feldspar peaks (K-feldspar and plagioclase) are seen between 27.5 and 28 2 theta in all but the un-reacted proppant. Particular attention was paid to the white, fine-grained material produced by experiment 1. This material was concentrated and analyzed separately. Figure 7b is an X-ray diffraction pattern of the white mineral aggregate products in experiment 1. Figure 7c shows diffraction patterns of the clay sized fraction (< 5 micrometers) of the same material. The dominant sheet silicate identified in the X-ray pattern is chlorite, which is accompanied by a trace of illite.

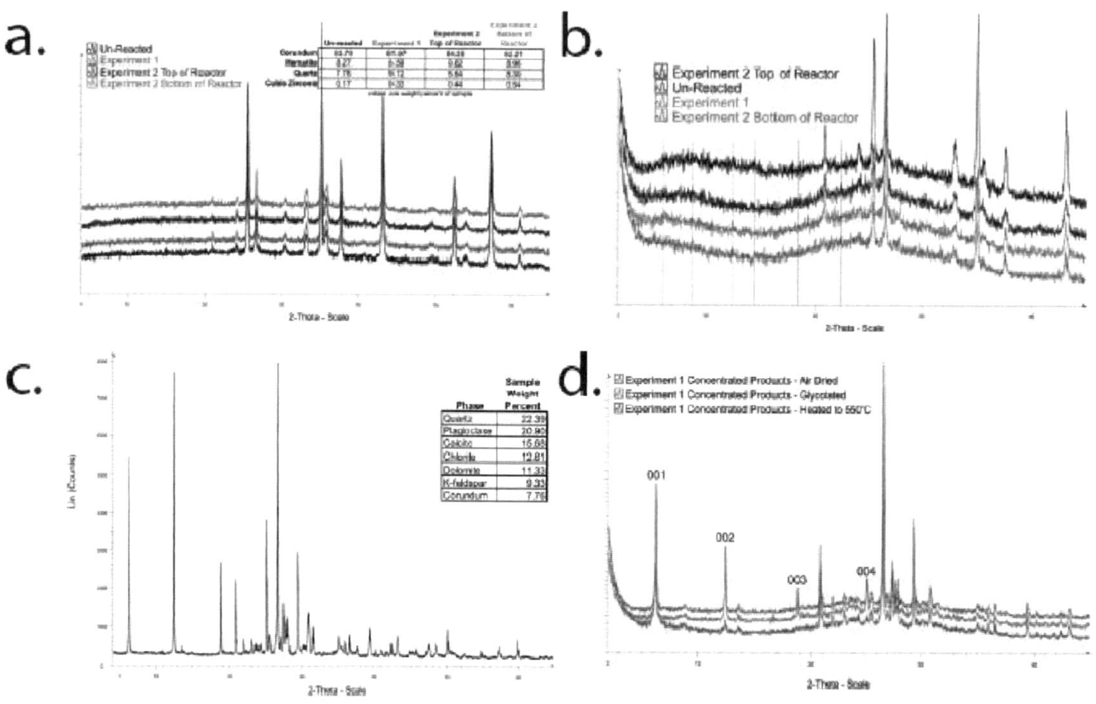

Figure 7a-d. (a). X-ray diffraction patterns of the un-reacted and reacted bulk proppant. (b). X-ray diffraction patterns of the glycolated clay sized fraction (<5 micrometers) of the un-reacted and reacted proppants. Vertical solid lines have been drawn at the primary peak locations of the bauxite minerals, boehmite (14.5 2 theta), gibbsite (18.3 2 theta), and diaspore (22.3 2 theta). A trace of boehmite is present in the sample 'Experiment 2 Top of Reactor', likewise a trace of gibbsite is present in the 'Un-Reacted' sample. Vertical dotted lines show the peak locations of the chlorite 001 and 002 peaks, a trace of which can be seen in sample 'Experiment 1'. The vertical dashed line shows the location of the 001 illite peak, which is present in trace amounts in sample 'Experiment 2 Bottom of Reactor'. (c). X-ray diffraction pattern and composition of the fine-grained mineral aggregate produced in experiment 1. It should be noted that some proppant is present in the sample, as indicated by the presence of corundum. (d). X-ray diffraction patterns of the clay sized fraction (<5 micrometers) of the material produced in experiment 1. Chlorite is the dominant sheet silicate ("clay" mineral) in the sample, with a trace of illite (001 at 8.8 2 theta, shown by vertical black line). The chlorite peaks are labeled 001, 002, etc. The large peak at 26.6° 2 theta is quartz; the broad hump centered at this angle is due to amorphous silica. There are no shifts between the air dried and glycolated patterns indicating there is no swelling clay in the sample. After heating to 550°C for one hour, the intensity of 001 peak increases greatly and shifts slightly (a few tenths of a degree 2 theta) and the 002, 003, & 004 reflections are significantly weakened; this is characteristic of chlorite.

QEMScan Analyses

QEMScan analysis of proppant from experiment 2 (Fig. 8) shows the mineralogical variation in the composition of the proppant itself, as well as the deposition of various minerals on the surface of the proppant. The proppant is mainly composed of Al and Fe oxides. The surface mineral deposits are identified as aluminum oxides and various silicates, including plagioclase, muscovite, quartz, and undifferentiated amorphous silicates. These mineral deposits appear to be layered (Fig. 8), likely as a function of their solubilities.

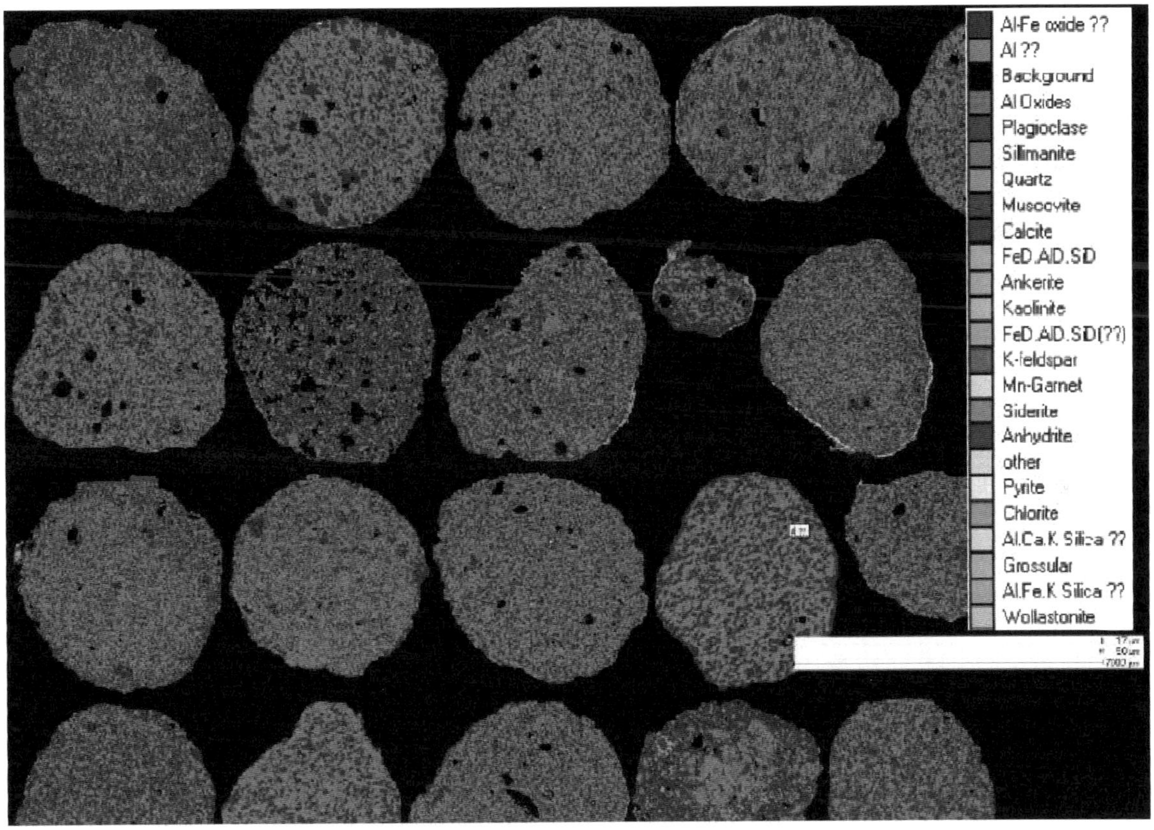

Figure 8. QEMScan analysis of proppant reacted in experiment 2.

Discussion

Mineral deposition on the proppant during experiment 2 is likely the result of the experimental methodology in removing the fluid from the apparatus. By allowing the fluid to cool within the reactor, minerals that were in equilibrium with the fluid at 200°C would have been at concentrations above their saturation limits as the fluid cooled. Due to this effect, the apparatus used for future experiments was modified to allow extraction of reactor liquids while preventing flashing and minimizing cooling time and associated precipitation. An additional valve at the bottom of the reactor will be used to remove reactor liquids. High pressure nitrogen will force liquids out through the bottom of the reactor. Fluids will flow down a length of tubing and through an ice bath to quickly cool the fluids. The fluids will then be collected and diluted to prevent mineral precipitation.

In order to more closely simulate in situ geothermal conditions, de-ionized water spiked with silica will be used for on-going experiments. This should minimize the dissolution of minerals from the proppant. The silica-spiked solution will be prepared using tetraethyl orthosilicate (TEOS) as the silica source. TEOS will be added to de-ionized water in an attempt to achieve a silica concentration of 265 ppm (the quartz saturation limit at 200°C). Granite will also be added to understand the chemical interactions between the proppant, rock, and fluid to more closely represent a geothermal system. Granite will be used as the rock type for on-going testing because it is an appropriate proxy for typical reservoir rocks at EGS sites. Several pieces of granitic core from Glacier Peak, Canada, a fossil geothermal system, will be used.

Conclusions

When using DI water at 200°C, the proppant has shown some signs of dissolution in these experiments. Changes to the chemistry of the fluid, as well as to the methodology of the experiments, should yield more realistic results as to the durability of bauxite proppants under geothermal conditions. The results of these experiments will be used as constraints in geochemical models to further understand these reactions and predict the chemical behavior of proppants in a reservoir environment.

Acknowledgements

This study is funded by the U.S. Department of Energy under grant **DE-FG36-08GO18189.**

References

Cooke Jr., C.E., 1977. "Fracturing with a high-strength proppant." Journal of Petroleum Technology, p. 1222-1226.

Kaszuba, J. P., Janecky, D. R., Snow, M. G., 2005. "Experimental evaluation of mixed fluid reactions between supercritical carbon dioxide and brine: Relevance to the integrity of a geologic carbon repository". *Chemical Geology*, v. 217, p. 277-293.

Kaszuba, J. P., Janecky, D. R., Snow, M. G., 2003. "Carbon dioxide reaction processes in a model brine aquifer at 2000 C and 200 bar: Implications for geologic sequestration of carbon." *Applied Geochemistry*, v. 18, p. 1065–1080.

Legarth, B., Tischner, T., Huenges, E., 2003. "Simulation experiments in sedimentary, low-enthalpy reservoirs for geothermal power generation, Germany." *Geothermics*, v. 32, p. 487-495.

Rosenbauer, R. J., Koksalan, T., Palandri, J.L., 2005. "Experimental investigation of CO_2-brine-rock interactions at elevated temperature and pressure: Implications for CO_2 sequestration in deep-saline aquifers." *Fuel Processing Technology*, v. 86, p. 1581–1597.

Seyfried Jr., Janecky, W.E., Berndt, D.R., 1987. "Rocking autoclaves for hydrothermal experiments, II. The flexible reaction cell system." In: Ulmer, G.C.; Barnes, H.L. (Eds.), Hydrothermal Experimental Techniques. John Wiley & Sons, p. 216-239.

Predicting Thermal Conductivity of Geothermal Reservoir Rocks

Dan Brinton, Joseph Moore, John McLennan, Clay Jones

Abstract

Understanding rock temperatures and stress fields will be important in predicting fracture behavior during enhanced geothermal reservoir development. For many geothermal reservoirs, thermophysical property data (coefficients of thermal expansion, thermal conductivity and specific heat) are limited. Thermal conductivity, diffusivity and specific heat are relevant for quantifying the power potential and forecast for prospective and operating geothermal plants and govern the way temperature gradients move through the reservoir. The coefficient of thermal expansion and certain other mechanical parameters of the reservoir such as Young's modulus and Poisson's ratio intrinsically contribute to thermal alteration of in-situ stresses in response to temperature perturbations during reservoir development by hydraulic fracturing and during heat production.

Recognizing the lack of a substantial database of thermophysical properties, standard geophysical well logs and mineralogic analysis of core and cuttings from the Raft River Geothermal Field have been used to predict thermal conductivity. Predictions are compared with the experimentally-determined values from core samples collected from Raft River Geothermal well RRG-3C.

The work will support ongoing investigations to thermally and hydraulically stimulate well RRG-9 under a US DOE sponsored grant. RRG-9 was drilled to 6089 feet measured depth (MD) in the Raft River Geothermal Field located in southern Idaho, near the Idaho-Utah state line. The field lies within the Basin and Range province. In the 1970s to early 1980s, the field was the site of intense study by the US Department of Energy. During that period a significant amount of geologic and reservoir data were collected. The reservoir is developed in Precambrian metamorphic and granitic rocks beneath approximately 5000 ft of Tertiary volcaniclastic deposits. U.S. Geothermal, Inc. operates a 13-MW plant at the site.

Introduction

The Raft River valley is located in southern Idaho, just north of the Utah border. Geothermal wells have been drilled in the valley to depths of up to 6500 feet and have intersected three distinct geologic formations. The Precambrian basement is comprised of quartz monzonite which is overlain by a series of four metasedimentary units: The Older Schist, the Elba Quartzite, the Upper Narrows Schist and the Quartzite of Yost (Blackett & Kolesar, 1983). Rock units from the Precambrian basement have been correlated with rocks in the Raft River Range (Covington, 1980). Above these basement rocks lies a thick (up to 5250 feet) series of highly fractured tuffs, sandstones and siltstones known as the Salt Lake Formation. The Salt Lake Formation is overlain by the Raft River Formation, comprised of sand, silt and gravel; the boundary between the Salt Lake and Raft River formations is gradational (Blackett & Kolesar, 1983).

Two major north-south trending fault zones—the Bridge Fault and the Horse Well Fault—have been identified on the east side of the valley in the Tertiary sediments (Blackett & Kolesar, 1983). These fault zones are responsible for the upward transport of the geothermal water from greater depths. In addition to the Bridge and Horse Well faults, there is some evidence, including lateral displacement of volcanic rocks at the surface (Blackett & Kolesar, 1983), for a shear fault in the Precambrian basement known as the Narrows structure.

Meteoric recharge of the hydrothermal system likely occurs in the Raft River and Albion mountains (Covington, 1980). Water is circulated at depth in the fractured basement, flows along the Narrows structure and then flows up the Bridge and Horse Well faults to the surface (Covington, 1980). The majority of the hydrothermal activity in the valley coincides with the proposed intersection of the Bridge and Horse Well fault zones with the Narrows structure.

Nine wells have been drilled to exploit the Raft River geothermal resource. Raft River Geothermal well 9 (RRG-9) is currently used as an injection well. A thermo-hydraulic fracture treatment campaign is scheduled to be conducted in late 2010 or early 2011. Because of the temperature difference between cold injection fluids and hot reservoir rocks, thermal fracturing, in addition to hydraulic fracturing, is expected to occur over the life of the reservoir. This thermal fracturing is anticipated to further increase the total fracture surface area beyond the surface area resulting from hydraulic fracturing alone. Perkins and Gonzalez suggested that these thermal fractures may extend perpendicular to the main fracture face creating a fracture network (1985).

The extent to which thermal fracturing occurs and makes a significant contribution to the total fracture surface area of the geothermal reservoir will be controlled by the thermophysical properties of the rock, existing fractures and the in-situ stress regime. The thermal conductivity, specific heat and density of the reservoir rocks will determine the rate of cooling in the reservoir. The coefficient of thermal expansion of the reservoir rock will control stress changes in the cooled region. Thermophysical properties such as thermal conductivity, specific heat and coefficient of thermal expansion are important in modeling and predicting fracture shearing and propagation in geothermal reservoirs. Laboratory-obtained values of these key properties have been limited by the scarcity of geothermal core. Standard geophysical well logs and mineralogic composition data from analyses of core and cuttings can be used to predict thermal conductivity.

Thermal Fracturing

During exploitation of a geothermal reservoir, thermal energy is removed from the reservoir by cool water flowing through the rock in the case of high permeability hydrothermal reservoirs, or along fracture faces in the case of enhanced geothermal systems where matrix permeability is assumed to be small. Heat extraction results in an ever-growing cooled region. As the cooled region continues to grow, the temperature gradients in the reservoir become smaller until it becomes uneconomical to operate the well.

In an enhanced geothermal system, the surface area available for heat transfer is limited by the surface area of the relatively few natural and engineered fractures in the reservoir formation. The rate of thermal drawdown in an enhanced geothermal system would be significantly greater than in a traditional permeable hydrothermal reservoir. Bažant et al. suggested that thermal fracturing associated with reservoir cooling may increase the effective size of the reservoir by opening up new cracks for fluid flow and heat transfer (1979). As the surfaces of a fluid channels cool, shrinkage cracks will begin to form normal to the main fluid channel. Bažant et al. predicted that water flux through these shrinkage cracks could be expected to be on the same order of magnitude as water flux through the main cracks (1979). Perkins and Gonzalez addressed the same phenomenon in hydraulic fracture applications in hot, permeable, convection dominated oil reservoirs from the perspective of modification of in- situ stresses due to rock shrinkage (1985).

It is evident that in low matrix permeability fractured geothermal reservoirs that heat conduction controls thermal output and depletion. The thermal conductivity, heat capacity and density of a substance govern the way heat flows through the substance. As applied to reservoir management and thermal fracturing, it will be important to be able to quantify these key properties in order to predict the way transient temperature profiles will move through the reservoir.

Predicting Thermal Conductivity

Recognizing the importance of knowing the reservoir properties such as thermal conductivity, as well as the difficulty and expense associated with an extensive coring and laboratory testing campaign, the present work outlines a method for obtaining the reservoir thermal conductivity from geophysical well logs and phonon conduction theory. This method was first described by Williams and Anderson (1990). Various methods of predicting rock thermal conductivity have been outlined, including prediction directly from geophysical logs or from rock compositional data; however these methods are strictly empirical and largely ignore the physics of heat transfer in solids. Williams and Anderson, however, combined easily-obtained well log data with the theory of phonon conduction in order to produce a model which can be applied generally (1990).

Heat is conducted in non-metallic solids primarily by crystal lattice vibration (Berman, 1976). This is known as phonon conduction. Energy can be stored and transferred in solids in the form of atoms vibrating relative to their equilibrium position (Yang, 2004). In the presence of a temperature gradient, lattice waves known as phonons travel down the gradient from regions of greater vibration, associated with elevated temperature, to regions of lesser vibration (Berman, 1976; Yang, 2004). The thermal conductivity of a crystal is a function of the heat capacity of the crystal, the phonon velocity and the mean free path (Yang, 2004). In non-porous rocks, then, thermal conductivity is a complex function of composition, crystal geometry and temperature. In porous rocks, prediction of thermal conductivity is expected to be further complicated by pore geometry, the presence of pore fluids and rock-fluid interactions.

Williams and Anderson presented a semi-empirical model of thermal conductivity in low-porosity rocks. Their model was based on geophysical well log responses, phonon conduction theory and laboratory testing of aggregate minerals. The equation for phonon conduction presented by Williams and Anderson is given as Equation 1

$$\lambda_{phonon} = 0.7531 + 0.1005\left(\frac{aV_m\mu^2}{3K_ST}\right) \qquad 1$$

where a is the mean interatomic distance, V_m is the mean phonon velocity, μ is the shear modulus, K_s is the bulk modulus and T is the temperature (Williams & Anderson, 1990). The temperature in Equation 1 is the temperature at which the experimental data reported by Williams and Anderson was obtained: 298 K. When this relationship was first presented, Williams and Anderson derived the shear and bulk moduli from compressional and shear wave sonic velocities and the mean interatomic distance was inferred from an approximation which used the photoelectric factor obtained from the photoelectric log. In the current study of Raft River geothermal wells, neither shear wave velocity nor photoelectric factor logs were available. Alternate methods for approximating shear wave sonic velocity and photoelectric factor have been used.

Brocher presented an empirical relationship between compression and shear wave velocities based on a large data set which included data from Quaternary alluvium and Salinian granites (2005). "Brocher's regression fit" is given as Equation 2.

$$V_S = 0.7858 - 1.2344V_P + 0.7949V_P^2 - 0.1238V_P^3 + 0.0064V_P^4 \qquad 2$$

The mineralogy of core and cuttings samples from well RRG-3C and RRG-3D were determined using X-ray diffraction spectroscopy (XRD). Approximate elemental compositions were determined from the mineralogy. The photoelectric factor for each mineral was calculated from the approximate elemental composition as the sum of the products of the atomic weights, photoelectric cross sections and numbers of each atom in the mineral, all divided by the molecular weight. Calculated photoelectric factors were compared with photoelectric factors tabulated by Hashmy and Alberty (1992). Table 1 compares the photoelectric factors obtained by the two methods. The photoelectric factors for each sample were then calculated as the average of the photoelectric factors of the minerals weighted by their mass fractions.

A principal component analysis (PCA) was conducted on well log data from Raft River well RRG-3 over the interval from approximately 4250 to 5800 feet MD. The PCA was performed on neutron porosity, gamma, formation density and compressional wave slowness logs along with a log of the difference between the neutron porosity and density porosity logs. PCA is a multivariate data analysis method which allows multivariate data sets to be reduced to a smaller number of variables or principal components. This is accomplished by performing an eigenvector decomposition of the variance-covariance matrix. The resulting eigenvectors and eigenvalues define the principal components of the data set.

Any data point can be projected onto the principal component just as, in two dimensions, any point can be projected onto the x or y axes. The projection of a data point onto the principal component is called the *score* of the data point. Principal component scores have been correlated with the photoelectric factor from analysis of core and cuttings. A linear relationship (Figure 1; Equation 3) emerged between the principal component scores and the photoelectric factor. While the goodness of fit is not ideal, an analysis of variance indicates that the linear relationship is statistically significant. The calculated value of the F statistic is 8.66 which is greater than the critical value of the F statistic (F=5.12) at the 95% confidence level.

Table 1 - Photoelectric factors of minerals present in core and cuttings from well RRG-3

Mineral	Photoelectric Factor	
	Calculated	Tabulated[a]
Biotite	12.39	6.27
Calcite	5.09	5.08
Chlorite	3.51[b]	6.3
Chlorite/Smectite	5.61[c]	
Epidote	6.87	
Fluorite	6.56	6.67
Hematite	21.94	21.48
Illite	2.44[d]	3.45
K-feldspar	2.89	2.86
Kaolinite	1.52	1.83
Microcline	2.89	2.86
Muscovite	2.44	2.4
Orthoclase	2.89	2.86
Plagioclase[e]	1.76[f]	1.68[f]
Albite	1.76	1.68
Anorthite	3.06	3.13
Pyrite	17.42	16.97
Quartz	1.81	1.81
Titanite	22	
Zeolite[e]	1.55[g]	
Analcime	1.55	
Chabazite	2.43	

a. from Hashmy and Alberty (1992)
b. as clinochlore
c. as corrensite
d. as muscovite
e. undifferentiated
f. as albite
g. as analcime

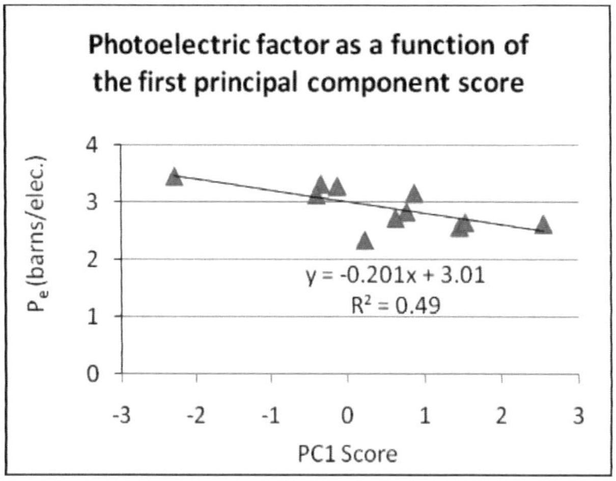

Figure 1 - A linear correlation between photoelectric factor (y axis) and principal component score (x axis) was found. This relationship allows for the determination of photoelectric factor from well logs.

$$P_e = 3.01 - 0.201 \cdot (PC1Score); \quad R^2=0.49 \qquad 3$$

The mean interatomic distance has been calculated as prescribed by Williams and Anderson: The effective atomic number has been calculated as (Ellis, 1987)

$$Z = 10P_e^{0.2778} \qquad 4$$

The mean atomic mass has been calculated as described by Williams (1989)

$$\overline{M} = \frac{(Z + 26.0)}{1.88} \qquad 5$$

Finally, the mean interatomic distance has been calculated as

$$a = \left(\frac{\overline{M}}{\rho}\right)^{1/3} \qquad 6$$

Thermal conductivity has been calculated for the rock solids over the entire depth range by the method of Williams and Anderson (Equation 1).

It is worth noting that the equation put forth by Williams and Anderson is a strong function of sonic wave velocities through both the mean phonon velocity and the bulk moduli. Thermal conductivity determined using this method is affected by the presence of alteration and microfracturing (Williams & Anderson, 1990). This was reemphasized by Pribnow and Sass (1995). They also pointed out that the Williams/Anderson model does not account for the effect of shear wave splitting associated with anisotropic rocks. The model then is most likely to be accurate in isotropic or flat-lying anisotropic rocks (Pribnow & Sass, 1995). Additionally, because the sonic tool analyzes waves traveling vertically, the thermal conductivity from the Williams/Anderson method is the *vertical* thermal conductivity (Williams & Anderson, 1990). While still valuable, for the present application the quantity of most interest is the *horizontal* thermal conductivity.

Experimental

Laboratory testing has been conducted on two core plugs from Raft River well RRG-3C to determine thermal conductivity. The core sample plugs were saturated and loaded to approximate in-situ conditions. The thermal conductivity of the two samples at various temperatures is reported in Table 2.

To allow comparison between the model of Williams and Anderson and the results of laboratory experiments, the experimental thermal conductivities have been extrapolated to 25°C, the temperature at which the Williams/Anderson experiments were conducted. Table 3 compares extrapolated experimental thermal conductivities with those predicted by the model.

Conclusions

The photoelectric factor of reservoir rocks shows a linear relationship with the principal component score derived from a principal component analysis of the neutron porosity, gamma, formation density, compression wave slowness and porosity difference logs of Raft River well

RRG-3. Results of the laboratory thermal conductivity testing of plugs from core taken from Raft River well RRG-3 show good correlation with conductivity determined using principal component analysis and the method of Williams and Anderson for the sample taken from 5272 feet. The discrepancy observed for the sample taken from 5270 feet may be due to anisotropy. Additional experimentally-determined results are necessary to draw further conclusions.

Table 2 - Thermal conductivity of core from Raft River well RRG-3C. Samples were saturated with slightly saline water and loaded to approximate in situ conditions.

Sample ID	Depth (ft)	Saturant	Confining Pressure (psi)	Pore Pressure (psi)	Average Test Temperature (°C)	Average Thermal Conductivity (W/m-K)
EGM1-1	5270.1	0.4% NaCl	2120	100	34.0	1.775
					68.0	1.850
					138.0	1.929
					Average over Temperatures	1.815
EGM2-5	5272.15	0.4% NaCl	2120	100	33.4	1.210
					69.6	1.307
					133.5	1.315
					Average over Temperatures	1.277

Table 3 - Comparison between thermal conductivity determined by laboratory analysis and by the method of Williams and Anderson

Depth (ft)	Sample Number	Extrapolated Laboratory Thermal Conductivity (W/m-K)	Thermal Conductivity from method of Williams and Anderson (W/m-K)
5270	EGM1-1	1.755	1.293
5272	EGM2-5	1.187	1.194

Acknowledgements

The authors wish to thank the management of U.S. Geothermal, Inc. for providing drill cuttings for analysis. This work has been supported by the United States Department of Energy grant DE-EE0000215.

References

Bažant, Z. P., Ohtsubo, H., and Aoh, K. (1979). Stability and Post-critical Growth of a System of Cooling or Shrinkage Cracks. *International Journal of Fracture , 15* (5), 443-456.

Berman, R. (1976). *Thermal Conduction in Solids.* Oxford: Clarendon Press.

Blackett, R. E., and Kolesar, P. T. (1983). Geology and Alteration of the Raft River Geothermal System, Idaho. *Transactions of the Geothermal Resources Council,* (pp. 123-127).

Brocher, T. M. (2005). Empirical Relations between Elastic Wavespeeds and Density in the Earth's Crust. *Bulletin of the Seismological Society of America , 95* (6), 2081-2092.

Covington, H. R. (1980). Subsurface Geology of the Raft River Geotheremal Area, Idaho. *Transactions of the Geothermal Resources Council,* (pp. 113-115).

Ellis, D. V. (1987). *Well Logging for Earth Scientists.* New York: Elsevier.

Hashmy, K., and Alberty, M. (1992). Difficult Lithologies. In D. Morton-Thompson, & A. M. Woods, *Development Geology Reference Manual.*

Perkins, T. K., and Gonzalez, J. A. (1985). The Effect of Thermoelastic Stresses on Injection Well Fracturing. *Society of Petroleum Engineers Journal ,* 78-88.

Pribnow, D. F., and Sass, J. H. (1995). Determination of Thermal Conductivity for Deep Boreholes. *Journal of Geophysical Research , 100* (B6), 9981-9994.

Williams, C. F. (1989). Heat and Fluid Flow Within the Earth's Crust: In Situ Measurements and Mathematical Models. New York: Columbia University.

Williams, C. F., and Anderson, R. N. (1990). Thermophysical Properties of the Earth's Crust: In Situ Measurements From Continental and Ocean Drilling. *Journal of Geophysical Research , 95* (B6), 9209-9236.

Yang, J. (2004). Theory of Thermal Conductivity. In T. M. Tritt, *Thermal Conductivity: Theory, Properties, and Applications* (pp. 1-20). New York: Kluwer Academic/Plenum Publishers.

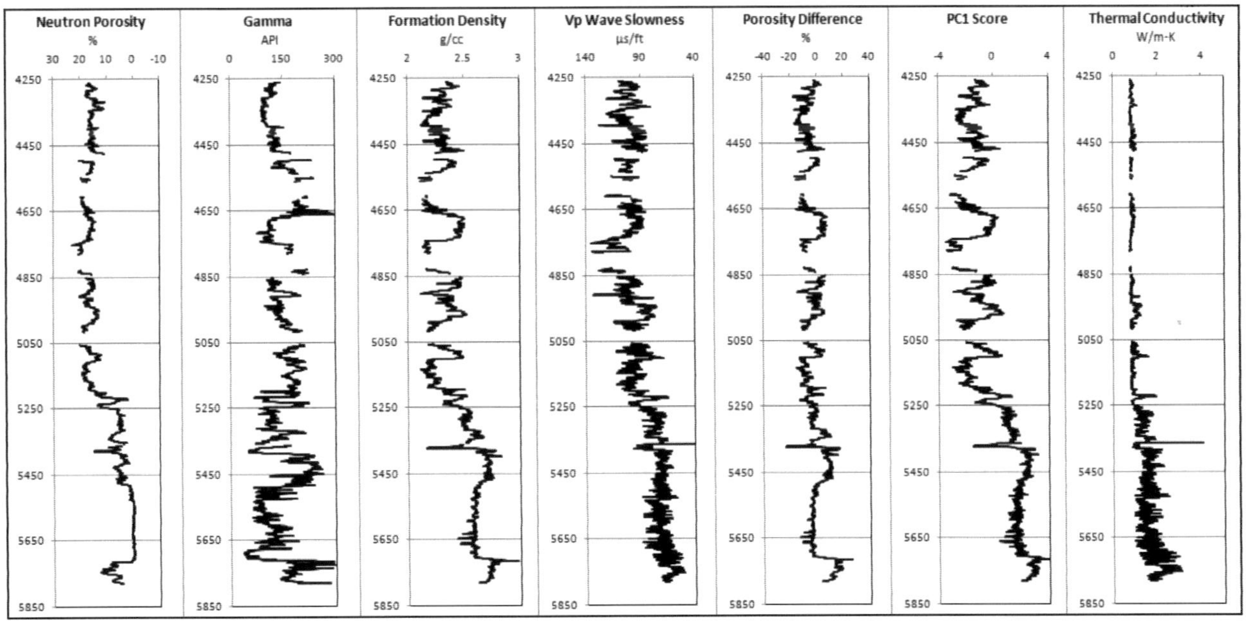

Figure 2 - Neutron porosity, gamma, formation density, and compressional wave slowness geophysical well logs. Also a log of the difference between neutron porosity and density porosity logs. These logs were used in a principal component analysis. Logs of Principal Component 1 Score and Calculated Thermal Conductivity are also shown. Thermal conductivity was calculated using the Williams/Anderson model.

(246a) Modeling and Simulation of Subsurface Processes in Oil Recovery and Carbon Storage

Angelo Lucia, University of Rhode Island, Kingston, RI and George A. Zyvoloski, Earth & Environmental Sciences Division, Los Alamos National Laboratory, Los Alamos, NM

Subsurface flow processes are important in oil recovery from shales and tar sands, carbon storage, and the movement of contaminants in ground water basins and typically involve complex physical phenomena that span several time and length scales. These processes are often modeled using coupled mass and energy balance equations along with phase equilibrium relationships to describe multi-phase flows through porous media. Simulation of steady or unsteady-state behavior of subsurface processes, reservoir simulation, includes a wide class of applications where both accurate phase densities and accurate phase equilibrium are essential. In addition, phase stability and phase equilibrium (or multi-phase flash computations) need to be performed quickly and reliably because they often consume a large fraction of the total reservoir simulation time (anywhere from 40 to 50%).

In this talk we describe a novel multi-scale modeling and simulation approach to subsurface processes. The proposed approach combines information from three time and length scales – the molecular scale, the bulk fluid phase scale, and the reservoir scale. At the molecular length scale we use coarse-grained NTP Monte Carlo simulations to determine internal energies of departure, UD, and create look-up tables of UD as a function of temperature, pressure and composition. This molecular scale information is communicated to the bulk phase length scale via look-up tables and interpolating formulae in order to predict the energy parameter in the new, multi-scale Gibbs-Helmholtz constrained (GHC) cubic equation of state (EOS) developed by Lucia (2009). The GHC EOS is used to compute fluid phase densities and component fugacity coefficients and solution fugacities which, in turn, are used to determine the number and types of phases as well as their corresponding compositions using Gibbs free energy minimization with full analytical partial derivatives. These minimum Gibbs free energy-based multi-phase flash computations are coupled to the reservoir simulator Finite Element Heat & Mass Transfer (FEHM) primarily developed and evolved by Zyvoloski (1975-present) to determine spatial and temporal distributions of vapor and liquid phases and multi-phase flows throughout the reservoir. FEHM uses a control volume finite element formulation of coupled heat and mass balance equations, Newton's method with analytical partial derivatives to solve the nonlinear algebraic equations that come from a discrete representations of the governing partial differential equations, and a pre-conditioned Krylov subspace method (partial LU factorization with GMRES) to solve the linear equations at each Newton step.

A number of interesting numerical examples are presented to demonstrate the efficacy of the proposed multi-scale modeling and simulation approach. We present comparisons of the GHC and SRK equations in the context of reservoir simulations. We also show that the multi-phase flash computations can be streamlined without loss of accuracy to provide very fast and reliable determination of phase behavior, can be readily interfaced with FEHM, and that coupling the GHC EOS with FEHM permits larger integration time steps and reliable simulations covering time horizons of 20 to 100 years.

See more of this Session: Simulation of Complex Subsurface Processes I
See more of this Group/Topical: Topical D: Chemical Engineering in Oil and Gas Production and Other Complex Subsurface Processes

New research simulation framework for flow modeling of complex physical system

Denis Voskov

Introduction

One of the most challenging type of a modern oil recovery processes is a compositional simulations. The main complexity of it is due to the nonlinear coupling of a multi-component multiphase flow and a thermodynamical phase behavior problem. An Equation of State (EoS) model usually employ to describe the phase partitioning for a mixture [6, 3]. For a given temperature, pressure, and overall composition of each component in the mixture, EoS computations define the phase state and compositions for each phase. That usually leads to the significant grow of computational time since EoS computations (that is an iterative process itself) needs to be performed for each computational cell and on each nonlinear iteration.

For a general purpose modeling, the different time-approximation scheme can be unify in a single simulation framework [4, 15]. Since governing equations, described multi-component flow coupled with thermodynamical equilibrium, are highly nonlinear, a Newton-Raphson method is used for a linearization. Base variables are usually divided into primary and secondary set based on alignment with equations. However, there is no single opinion which variables should be used as a base variables for such kind of linearization.

One of the most used formulations is the natural variables, when pressure, saturations, and phase mole fractions are used as the base set of unknowns (see [6]). One of the essential part of the natural-variables formulation is the variable substitution technique when we should use different combination of variables and equations (conservation equations and thermodynamical constraints) for a different phase state. Another widely used formulation is a different variations of molar variables [14, 5]. Here pressure and overall mole fractions used as a primary variables. Secondary variable set can be different for different formulations. This formulations look attractive since it is possible to avoid variable substitution when phase status is changing.

The phase behavior of multi-component mixture with arbitrary number of phases is another challenging problem. Several methods have been developed to improve the correctness of the phase stability results (see [18, 19]). However, all of these methods are based on global optimization techniques, computationally expensive and not guaranteed the convergence to the right solution. In the proposed simulation framework we use stability test based on adaptive tie-simplex parameterization (see [11], [20]) that improve both robustness and efficiency of phase behavior computations for multiphase multi-component simulation.

After a brief overview of mathematical statement for the problem, we are going to show the behavior of several nonlinear formulations and solution strategies for two-phase isothermal compositional simulation. Next the extension of two-phase isothermal framework to the flow of an arbitrary number of phases is presented. This extension helps to model a very complex processes such as thermal steam injection, low temperature CO2 injection to hydrocarbon reservoirs, flow in the gas-hydrates systems etc. Several representative examples are shown.

Conservation equations and constraints

In this section a general description of the governing equations for a multiphase flow in porous media is given. The transport equations for a system containing n_c components and n_p phases can be written as:

$$F_i = \frac{\partial}{\partial t}\left(\phi \sum_j x_{ij}\rho_j S_j\right) \quad (1)$$

$$+ \operatorname{div} \sum_j x_{ij}\rho_j \mathbf{u}_j + \sum_j x_{ij}\rho_j q_j = 0, \quad i = 1, \ldots, n_c. \quad (2)$$

Here ϕ denotes porosity, ρ_j is the density of phase j, S_j is a phase saturation and x_{ij} is the composition of component, i, in phase, j. All relations are functions depended on base (nonlinear) unknowns. These unknowns can be almost arbitrary chosen based on the nonlinear formulation. Below we present several formulations for isothermal compositional problem.

For description of flow for any phase we use Darcy's law:

$$\mathbf{u}_j = \left(\mathbf{k}\frac{k_{rj}}{\mu_j}(\nabla p - \gamma_j \nabla D)\right), \quad j = 1, \ldots, n_p, \quad (3)$$

where k_{rj} is the relative permeability of phase j, μ_j is the viscosity and k is the absolute permeability tensor.

Another assumption usually used for modeling of large scale flow in porous media is the thermodynamical equilibrium for components. In order to satisfy this condition we need to add $n_c \times (n_p - 1)$ equations define the equilibrium of chemical potential or fugacities for each component in each phase:

$$\psi_{ij}(p, T, x_{ij}) - \psi_{i,k}(p, T, x_{ik}) = 0, \quad \forall j \neq k \in \{1, \ldots, n_p\}, \quad i \in [1, \ldots n_c]. \quad (4)$$

Here p and T denote pressure and temperature, $\psi_{ij}(p, T, x_{ij})$ is the fugacity of component i in phase j.

Depends on the choice of formulation, different constraints can be added to the system in order to provide a closed mathematical statement. The phase behavior is an important type of computation needs to complete numerical framework and we describe it in the next section.

Phase behavior

For the phase behavior of a multi-component multiphase flow in large scale porous media we use the assumption of spontaneous thermodynamic equilibrium for each component in each phase, described by (4). This equilibrium condition can be used in combination with other equations to form a closed mathematical statement of problem.

Consistently with two-phase compositional framework we are going to use a two-stage procedure for a computation of phase equilibrium. First, the number of existing phases in each cell is determined and number of independent variables is defined based on the cell status. Next, if the status is changed, we need to run flash calculations to determine the compositions of the existing phases, x_{ij}.

Various methods can be used for that type of computations. First stage, called stability test, is the challenging procedure for system with arbitrary number of phases since a lot of local minima in solution of equilibrium problem exist. The second stage can be done using multiphase flash procedure, described in [17]. Since both stages of phase behavior computation is the second order high dimensional optimization it is usually a very computationally expensive procedure.

Here we propose a different approach for a phase behavior computation. We are going to use the fact that phase behavior is a continuous function of pressure, temperature and overall composition. (see [23]). That give us an opportunity to use Compositional Space Adaptive Tabulation, developed for two-phase systems ([10], [11]), in order to parameterize tie-simplex associated with phase behavior for discrete points in compositional space and for discreet values of pressure and temperature.

The CSAT framework allows us to use a generalized negative-flash computations in order to define the number of existing phases and corresponding phase composition. In [24] we demonstrate that single multi-phase negative flash could be not enough for accurate phase state identification and suggested multi-stage negative flash in order to cover this complexity.

We can use pressure and temperature defined for a given grid cell and build interpolation for each tie-simplex (e.g. tie-triangle for three-phase system) in the table. Then we can use closest tie-simplex to identify phase state for a given grid cell. There is strong theoretical and practical evidence ([20]) that we need only limited number of such tie-simplex tables to cover the whole range of composition associated with multi-component multiphase displacement process.

Numerical framework

The simulation framework for the research purposes should be highly configurable in order to check different scientific conceptions and ideas. One of the important part of any simulation software is the nonlinear formulation that includes a set of equations described the problem and base unknowns used for linearization of problem. Here we use numerical implementation based on Automatic Differentiation Expression Templates Library ADETL described in [13].

In the framework we specified different phase states for the simulation region (grid-block) that is usually closely related on thermodynamical state (number of

Table 1: Phase status table for three-phase system

Status	Number of phases	Array of existing phases		
		Phase 1	Phase 2	Phase 3
1	1	x	-	-
2	1	-	x	-
3	2	x	x	-
4	1	-	-	x
5	2	x	-	x
6	2	-	x	x
7	3	x	x	x

existing phases). Depending on the state we can use a different set of governing equations and nonlinear unknowns. The resulting Jacobian is constructed based on the specified variable set automatically since all the procedures used in residual construction are based on AD-type variables that allow to compute derivatives simultaneously with construction of analytical relation.

One of the important things is the possibility to construct and differentiate the all possible thermodynamical states corresponding to the system with specific number of co-existing phases. As we mentioned before, the negative-flash procedure estimate the number of existing phases based on the sign of phase fraction ν_j. Than we can define the status of phases based on the following relation:

$$N_s = \sum_{j=1}^{n_p} 2^{j-1} \delta_{(\nu_j > 0)} \tag{5}$$

Here $\delta_{(\nu_j>0)} = 1$ if $\nu_j > 0$ and 0 otherwise. The example of all possible states corresponding to three-phase system is presented in following table. It is possible to build a similar table for arbitrary number of phases. The phase status can be evaluated based on (5) and for the known number of status we can have the list of existing phases based on the table. That helps to unify the most of the computations based on a phase status for the particular grid-block.

Solution Methods

The problem we need to solve is highly nonlinear, and we use the Newton-Raphson method to linearize it. For each nonlinear iteration linear system contained Jacobian should be solved. In order to reduce the size of linear system, the complete set of equation is split into two parts. First part (primary equations) are the mass and energy conservation equations since they account for fluxes across cell boundaries and depend on the unknowns (variables) in neighboring gridblocks. The second part is the rest of the equations which are local for any discrete gridblock. The unknowns aligned with both parts are called primary and secondary variables. There is no special rules for alignment of equations and variables except the condition that both the primary and the secondary diagonal matrices should be invertible.

After the linear system is solved, we can update all the nonlinear variables. This update provides a nonlinear solution based on Newton-Raphson method. However, if we use this method blindly that can lead to inconsistency in nonlinear behavior since any Newton update is mathematical abstraction that could not fully preserve all the physics. In order to avoid nonphysical conditions and improve the nonlinear convergence, we introduce several corrections to the standard Newton update.

The time-stepping strategy based on convergence of nonlinear process in the L_∞ norm. Each successful timestep follow by increasing of the next timestep until its length reaches a limit. If the number of nonlinear iterations breaks a maximal number of iterations, current timestep should be reduced. Another reason for timestep to be reduced is the negative value of pressure on any Newton iteration or convergence problem of a linear solver.

Nonlinear compositional formulations

In this section we describe different nonlinear formulations for the isothermal compositional problem. In each formulation we are using different set of unknowns to solve the system described by (1) and (4) nonlinear equations. For each unknown we specified its dimension in square brackets. We also outline the sequence of steps used to compute the properties and the Jacobian matrix for each formulation.

Natural variables

For this formulation we are using following variables:

- p – pressure [1],
- S_1 – phase saturation [1],
- x_{ij} – phase composition of each component $[2n_c - 2]$.

In order to detect the phase change we use EoS computations. For single-phase gridblocks we need to run stability test to detect is the second phase can appear. In two-phase blocks we use S_1 to detect the phase disappearance.

In order to complete computational framework, we also need to express the molar phase fraction, ν_j, for two-phase cells as a function of saturation, namely,

$$\nu_j = \frac{S_j \rho_j}{\sum_k S_k \rho_k}. \tag{6}$$

Then the overall mole fraction of component i can be written as

$$z_i = \sum_j x_{ij} \nu_j. \tag{7}$$

For a single-phase mixtures, $\nu_j = S_j$ and $z_i \equiv x_{ij}$.

Molar variables

([5]) This formulation is a variant of the molar-variables formulation, where instead of the phase fraction $x_{i,1}$, equilibrium ratios, K_i, are used. In [12], use of $\ln K_i$ was suggested in order to make (4) more linear. The following unknowns are used:

- p – pressure [1],
- ν_1 – phase fraction [1],
- z_i – overall phase composition of each component $[n_c - 1]$,
- lnK_i – phase equilibrium ratio $[n_c - 1]$.

The phase appearance and disappearance is detected based on the similar to natural formulation rules except that to detect the phase disappearance ν_1 is used instead of S_1.

In order to complete computational framework, we need to define phase composition for two-phase state as

$$x_{i,2} = z_i / \left(1 + \nu_1(K_i - 1)\right) \tag{8}$$

$$x_{i,1} = K_i x_{i,2} \tag{9}$$

Then density of phase can be found from EoS and saturation computed from modification of (6).

Overall-composition variables

In this formulation, the following variables are used:

- p – pressure [1],
- z_i – overall composition of each component $[n_c]$.

For a given Newton iteration, a phase stability test is performed to determine if the overall composition, z_i, will split into two phases. If the gridblock is single phase, no additional computations are necessary. If a two-phase state is detected, flash calculations are performed to obtain the phase compositions $x_{i,j}$. The system $\mathbf{F} = 0$ is then solved for the given p, T, and z_i where:

$$F_i = f_{i,1}(p, T, x_{i,1}) - f_{i,2}(p, T, x_{i,2}), \quad i = 1, \ldots, n_c, \tag{10}$$

$$F_{i+n_c} = z_i - Vx_{i,1} - (1 - V)x_{i,2}, \quad i = 1, \ldots, n_c, \tag{11}$$

$$F_{2n_c+1} = \sum_{i=1}^{n_c} (x_{i,1} - x_{i,2}), \tag{12}$$

Note that derivatives of the phase fractions as a function of the base unknowns are required, which can be obtained as follows

$$\frac{\partial \mathbf{x}}{\partial \mathbf{z}} = \left(\frac{\partial \mathbf{F}}{\partial \mathbf{x}}\right)^{-1} \frac{\partial \mathbf{F}}{\partial \mathbf{z}}. \tag{13}$$

The rest of the computations can be done in a manner similar to the other mass (mole) formulations.

Volume-balance formulation

([1]) The main difference of volume-balance formulation for a fully-implicit simulation is the use of extensive molar variables, N_i, instead of intensive overall compositions, z_i. Usually N_i is defined as

$$N_i = \phi V \sum_{p=1}^{n_p} \rho_p S_p x_{i,p}. \tag{14}$$

In order to complete computational framework, we can compute $z_i = N_i / \sum_{i=1}^{n_c} N_i$. Following computations are similar to the molar formulation.

Isothermal two-phase examples

Here we present the numerical results for comparison of different nonlinear formulation described in previous section. For test we choose small 2D (20x20) reservoir with four high permeable channels: in the 3rd and 12th gridblock in the x-direction and in the 2nd and 8th gridblocks in the y-direction. The other parameters for the reservoir are

- the size of block is 100x100x1 meters
- the reservoir-matrix permeability is 1 D, the permeability of the channels is 20 D.
- porosity is 0.2
- relative permeability is a Corey function with gas exponents equal to three and oil exponents equal to two.
- production and injection wells located on opposite corners (1,1) and (20,20)

In this reservoir we tested three different compositional models

- 4 component, immiscible: initial oil is made of $\{C_1$ (20%), CO_2 (1%), C_4 (29%), C_{10} (50%)$\}$ at temperature $T = 100^oC$ and initial pressure of 80 bars (above the bubble point), injection gas is a mixture of two components $\{CO_2$ (90%),C_1 (10%)$\}$. The injection pressure is 140 bars, which is sufficiently above the dew point of the injected mixture; the production pressure is 50 bars, which is below the bubble-point pressure of the reservoir mixture.
- 4 component, miscible: the same initial mixture with injection gas $\{CO_2$ (90%),C_4 (10%)$\}$ and the same injection, production, and initial pressures. Note that injection pressure is higher than Minima Miscible Pressure (MMP) for this system that makes it miscible.
- 6 component, SPE5: system used for SPE 5 comparative study [7]. Gas is composed of $\{C_1$ (77%), C_3 (20%), C_6 (3%)$\}$. The oil composition is $\{C_1$ (50%), C_3 (3%), C_6 (7%), C_{10} (20%), C_{15} (15%)$\}$, C_{20} (5%)$\}$. Temperature of

Table 2: Test results for formulations comparison

N	Formulations	4 comp., immiscible		4 comp., miscible		6 comp. (SPE5)	
		Newton	EOS	Newton	EOS	Newton	EOS
1	Natural	373	5.02	379	6.17	247	8.10
	Molar	386	5.07	441	6.43	455 (45)	10.77
	Molar VB	403	5.07	484 (6)	6.92	403 (30)	10.12
	Overall	444 (62)	8.27	435 (47)	9.30	252	10.58
2	Natural	207 (29)	7.24	174	6.77	197 (43)	12.30
	Molar	213 (20)	6.79	268 (9)	7.99	535 (153)	13.49
	Molar VB	201 (2)	5.78	378 (43)	9.18	395 (67)	11.29
	Overall	323 (120)	12.01	271 (73)	12.27	259 (100)	20.77
3	Natural	206 (28)	7.69	174	8.36	181 (34)	11.55
	Molar	198 (20)	7.35	179	8.33	138	10.35
	Molar VB	203 (20)	7.31	220 (21)	8.73	136	10.71

reservoir is $T = 70^oC$, initial pressure is 210 bars, injection pressure is 340 bars and production pressure is 100 bars. In described condition the displacement process is multi-contact miscible.

Below you can see the table with results. Three types of simulation was done for all systems and nonlinear formulations:

1. Normal simulation with not-aggressive time step, corresponding CFL numbers for three cases are $CFL_{max} = 9.2$, 9.0 and 14.8 respectively;

2. Normal simulation with aggressive time step, corresponding CFL numbers for three cases are $CFL_{max} = 35.2$, 35.5 and 29.1 respectively;

3. Simulation with the flash reduction for aggressive time step, the CFL numbers are similar to previous.

In the last type of simulation we apply exact flash procedure and correct corresponding variables for any Newton iteration. The maximal number of Newton iterations allowed for each time step was $N_{max} = 20$.

For each test run you can see two numbers corresponding with an accumulative number of Newton iteration for each run, including number of wasted iteration in brackets and an average number of EoS iteration (either stability or flash) done for each Newton iteration in each grid block. For all the simulations number of linear iterations was quite similar and that's why we don't report it in the table. Since the amount of computations for each Newton iteration is almost constant, the number of Newton iterations really corresponds to the simulation performance. Of course, the number of EoS iterations affects the total time as well, but it can sufficiently reduces to almost nothing using CSAT approach [10, 11]. There is no results reported for the overall-composition formulation in last type of run because these numbers are identical to the second case since this formulation always requires exact flash for any Newton iteration.

As you can see, for the first type of simulation (lower time step limit), the number of Newton iterations is quite different. The formulation based on natural variables looks the most attractive among the rest for the full range of simulated problems, but the difference is not really big. The number of EOS iterations is almost constant for all formulations except overall-composition, since it needs more flash iterations for two-phase cells. This picture changes when aggressive time-step is applied. Now the natural formulation is a way more advanced. That happened since the nonlinear chopping of base variables in case of natural formulation deals with phase saturation directly. That helps to avoid oscillations along a fractional flow curve that seems to be a dominant nonlinearity for compositional simulation with an intensive phase change. At the same time, the formulation based on an overall-composition run exact flash for any nonlinear iteration in order to detect phase disappearance. For the non-fully-converged solution that can introduce fake phase change and leads to oscillation that grow the number of wasted iterations. The rest of a molar-variable-based formulation just suffers from a worse nonlinear convergence.

The performance of molar variable set change when exact flash is added for the Newton iteration. In this case, the difference between molar formulations and overall-composition formulation is that for both molar formulations we use phase fraction ν_1 to detect phase disappearance for two-phase gridblocks before the flash correction. If a two-phase state is still valid, we do an exact flash in order to correct phase composition x_{cp}. That avoids oscillation based on phase detection, that is typical for overall-composition formulation, and improves convergence of molar formulations. As you can see from the table, both molar formulations demonstrate the large improvement and become almost as efficient as the natural formulation. At the same time, the flash correction improves the natural formulation very slightly.

Examples of three-phase flow

Here we present several numerical examples of the displacement in the system with three EOS-based phase. The nonlinear formulation is based on an extension of natural variables. The nonlinear solver is the same simple chopping strategy. In all examples, the Stone 1 model is used to construct the oil relative permeability.

For a phase behavior computations we use the combination of the multiphase flash [17] with the stability analysis based on the Compositional Space Adaptive Tabulation (CSAT). For a three-phase system this algorithm includes parameterization of a three-phase region as well as all corresponding two-phase regions in the plane of the tie-triangle, corresponding to the three-phase region.

In [21] it was shown that it is not possible to enter the three-phase region from the one-phase region without a shock in the composition and this shock belongs to the extension of the tie-triangle plane. In [20] we show that the similar rule is valid for the system with an arbitrary number of phases. This fact helps to explain why CSAT approach is supposed to be effective even when a number of phases exceed two. We demonstrate this efficiency for two-phase isothermal [10, 11] and thermal [22] displacement problems and here we are going to demonstrate it for three-phase problems here.

All simulations are done based on 1D and 2D reservoirs models with following parameters:

- the size of block is 100x100x1 meters;
- the porosity is 0.2;
- the relative permeability is a Corey function with gas exponents equal to three and oil exponents equal to two;
- production and injection wells located on opposite sides of model (the first and the last blocks);
- the reservoir permeability is 1 D.

We run either 1D (100 grid blocks) or 2D (10×10 and 20×20 grid blocks) models. In 2D models we add two high-permeable (20 D) channels in order to provide a heterogeneity. We use the three-parameter Peng-Robinson Equation of State for a phase behavior model.

First type of the simulation is an injection of a steam-gas mixture in the heavy-oil reservoir represented by $\{C_1$ (5%), C_8 (10%), C_{20} (70%), H_2O (10%)$\}$. The injected steam-gas mixture is $\{C_1$ (20%), H_2O (80%)$\}$. Initial conditions are $p = 140$ and $T = 130^oC$, injection conditions are $p = 60$ bars.

As you can see in Fig. 1,a, in this case only one tie-triangle-plane is needed to parameterize an entire simulation composition that requires a phase identification. That happened because both the injection and initial compositions belongs to the almost single plane and all shocks between compositions in different phases belongs in this plane as well. Next we change injection composition and add the C_8 component, so the steam-gas mixture will become $\{C_1$ (15%), $\{C_8$ (5%), H_2O (80%)$\}$. In Fig. 1,b you can see that two tie-triangle-planes are parameterized by the CSAT method. Now two shocks that connects the injection and initial composition belong to the different tie-triangle-plane and we need both of them for a phase identification.

If we change the reservoir description and look at the displacement curve for the 2D homogenous reservoir (10×10 gridblocks), in Fig. 2,a you can see that the phase behavior for this case is quite similar. The solution is obviously more dispersive but we need only one more tie-triangle-plane to parameterize solution in this case. If we use better resolution (20×20 gridblocks), the number of parameterized tie-triangle-planes again drops to only two (see Fig. 2,b).

Next example is the injection of CO_2 into oil at cold conditions. The oil is modeling by $\{C_1$ (20%), C_2 (30%), C_6 (40%), CO_2 (10%)$\}$ and initial conditions are $p = 4$ bars and $T = -110^oC$. At this condition the second liquid hydrocarbon (CO_2-rich) phase appears in the system. We inject pure CO_2 at $p = 10$ bars and produce at $p = 2$ bars. In Fig. 3,a you can see solution of CO_2 displacement in 1D reservoir. Again, in order to describe phase behavior for the entire solution, CSAT parameterized only 5 tie-triangles-planes. And this number changes insignificantly

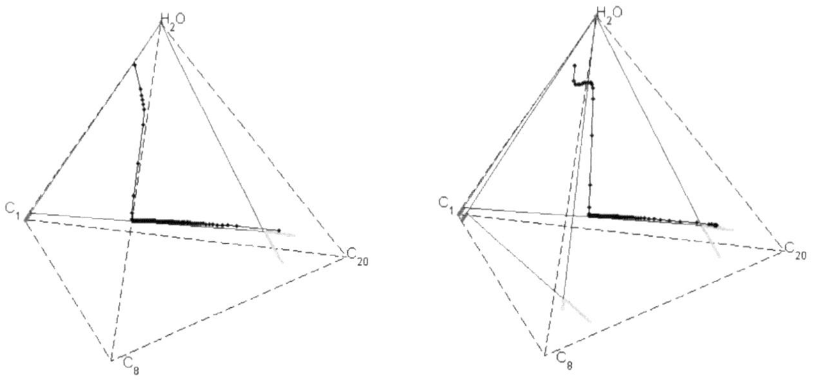

(a) (b)

Figure 1: Solution of oil displacement problem with steam-gas mixture for 1D homogenous reservoir. Tie-triangles with corresponding two-phase regions are presented in color.

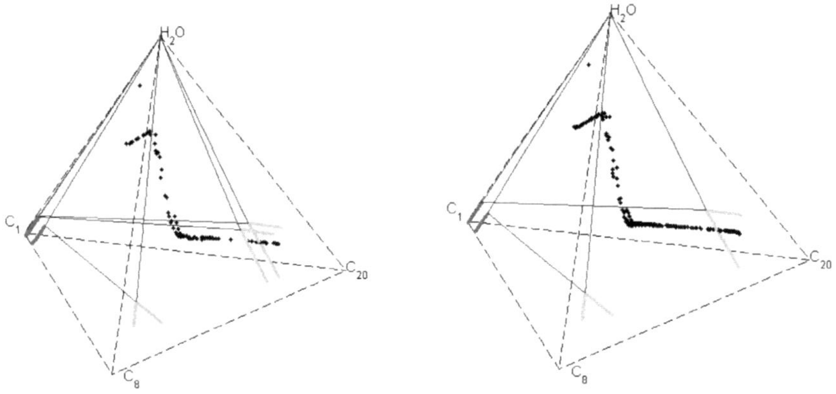

(a) (b)

Figure 2: Solution of oil displacement problem with steam-gas mixture for 2D heterogenous reservoir. Tie-triangles with corresponding two-phase regions are presented in color.

if we move from homogenous 1D model to heterogenous 2D (20×20 gridblocks) as seen in Fig. 3

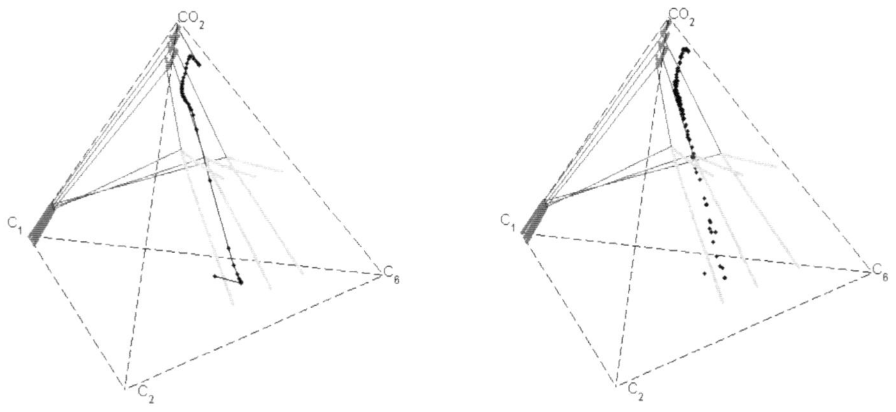

(a) (b)

Figure 3: Solution of oil displacement problem with pure CO_2 at cold conditions for 1D (a) and 2D (b) reservoirs. Tie-triangles with corresponding two-phase regions are presented in color.

Summary and Conclusions

The new framework for General Purpose Research Simulator (GPRS) based on an Automatic Differentiation with Expression Templates Library (ADETL) library is presented. This framework is used for various research related to the simulation of a complex physical problems associated with modern Enhancement Oil Recovery (EOR) processes.

The comparison of several widely used nonlinear formulations for isothermal compositional simulation is performed. The aim of this investigation was the nonlinear behavior associated with gas displacement processes in heterogenous reservoirs.

Several results of flow in porous media simulation for EoS-based three-phase system are also reported.

Acknowledgement

Thanks to Alireza Iranshahr, Rami Younis and Yifan Zhou for contribution to the simulation framework. Acknowledgements to the Chevron, Schlumberger and Total companies and the SUPRI-B Industrial Affiliates program on reservoir simulation at Stanford University for financial support.

References

[1] Act G., Doleshall S., Farkas E., General Purpose Compositional Model. - SPEJ, 1985, vol. 25, N 4.

[2] Aziz, K. and Settari, A., Petroleum Reservoir Simulation, Applied Science Publishers, London, England, 1979

[3] Aziz, K. and Wong, T.W., Considerations in the Development of Multipurpose Reservoir Simulation Models, proceedings of the 1st and 2nd International Forum on Reservoir Simulation, Alpbach, Austria, September 12-16, 1988 and September 4-8, 1989.

[4] Cao H., Development of Techniques for General Purpose Simulators, Ph.D. Thesis, Stanford University, 2002

[5] Chien M.C.H., Lee S.T., Chen W.H., A New Fully Implicit Compositional Simulator. - SPE 13385, 1985.

[6] Coats K.H., An Equation of State Compostional Model. - SPEJ, 1980, v.20, N 5.

[7] Killough J.E., Kossack C.A., Fifth comparative solution project: evaluation of miscible flood simulators, SPE 16000, 1987.

[8] Michelsen M.L., The isothermal flash problem: Part I. Stability. Fluid Phase Equilibria, 9: 1-19, 1982.

[9] Michelsen M.L., The isothermal flash problem: Part II. Phase-split calculation, Fluid Phase Equilibria, 9: 21-40, 1982.

[10] Voskov D., Tchelepi H., Compositional Space Parameterization: Theoretical Background and Application for Immiscible Displacements, SPE Journal 14 (3), 2009, (paper SPE 106029 presented at 2007 SPE Reservoir Simulation Symposium, Woodlands, TX, February 26-28).

[11] Voskov D., Tchelepi H., Compositional Space Parameterization: Multi-Contact Miscible Displacements and Extension to Multiple Phases, SPE Journal 14 (3), 2009, (paper SPE 113492 presented at 2008 SPE/DOE Symposium on Improved Oil Recovery, Tulsa, Oklahoma, April 20-23).

[12] Wang P., Yotov I., Wheeler M., Arbogast T., Dawson C., Parashar M., Sepehrnoori K.: A new generation EOS compositional reservoir simulator: Part 1 - formulation and discretization, SPE 37979, 1997

[13] Younis R.M., Aziz K., Parallel Automatically Differentiable Data-Types for Next-Generation Simulator Development, Proc. of Reservoir Simulation Symposium, Houston, TX, February, 2007.

[14] Young, L.C. and Stephenson, R.E., A Generalized Compositional Approach for Reservoir Simulation, SPEJ, October 1983, pp727-742.

[15] Voskov D.V., Younis R.M., Tchelepi H.A., General Nonlinear Solution Strategies for Multiphase Multi-component EoS Based Simulation, SPE 118996, presented at 2009 SPE Reservoir Simulation Symposium, Woodlands, TX, February 2-4.

[16] Chien, M.C.H., Yardumian, H.E., Chung E.Y. and Todd, W.W., The Formulation of a Thermal Simulation Model in a Vectorized, General Purpose Reservoir Simulator. - SPE 18418, 1998.

[17] Michelsen, M.L. and Mollerup, J.M., Thermodynamic Models: Fundamentals and Computational Aspects, Tie-Line Press, Holte, Denmark (2004).

[18] C.M. McDonald, C.A. Floudas, Global Optimization for the Phase Stability Problem, AIChE J., 41, 7 (1995) 1798-1814.

[19] D.V Nichita, S. Gomez, E. Luna, Phase stability analysis with cubic equations of state by using a global optimization method, Fluid Phase Equilib., 194-197 (2002) 411-437.

[20] Voskov D., Tchelepi H., Tie-simplex based mathematical framework for thermodynamical equilibrium computation of mixtures with an arbitrary number of phases, Fluid Phase Equilibria, 283, 1-2 (2009) 1-11.

[21] T. LaForce, K. Jessen, F. M. Orr, Four-component gas/water/oil displacements in one dimension: part II, example solutions, Transport in Porous Media, 72, 1, (2008) 83-96.

[22] Iranshahr A., Voskov D., Tchelepi H., Efficient and Robust Phase Equilibrium Computations for Thermal Compositional EoS Based Simulation, SPE Journal, 15, 2, (2010) 537-548.

[23] Iranshahr A., Voskov D., Tchelepi H., Compositional Space Parameterization for Compositional Reservoir Flow Simulation, AIChE Annual Meeting, Nashville, TN, November 8-13, 2009.

[24] Iranshahr A., Voskov D., Tchelepi H., A Generalized Negative Flash Procedure for Phase Equilibrium Computations Related to Carbon Dioxide Sequestration, AIChE Annual Meeting, Nashville, TN, November 8-13, 2009.

(246c) New Stochastic Approach to Construct Porous Media Geometry Models and Simulation of Non-Darcy Flows

Feng Xiao, Michael Newman and Xiaolong Yin, Petroleum Engineering, Colorado School of Mines, Golden, CO

We present a new method to construct two- and three-dimensional porous media geometry models from Voronoi diagrams. The parameters of the algorithm can be adjusted to produce isotropic and anisotropic fully percolated geometries with both low and high effective porosities (10% to 50%). The permeability calculated using D2Q9 and D3Q19 lattice Boltzmann methods with multiple relaxation time (MRT) collision operators matched well with existing correlations. The tortuosity of the models, calculated from diffusion of tracers, ranged from 1.2 to 1.5. As an application, we studied regime transition from Darcy to non-Darcy flows. In Darcy flows, the superficial velocity through the porous media is a linear function of applied pressure gradient; in non-Darcy flows, due to the inertia of the fluid, the superficial velocity becomes a nonlinear function of the applied pressure gradient. The observed variation in the apparent permeability with the Reynolds number agreed well with existing correlations. The impact of porous media texture on flow regime transition was investigated by randomly removing grains. Porous media geometries with grains removed showed transitions to non-Darcy flows at lower Reynolds numbers and the deviations had higher slopes than those without removed grains at a similar porosity. Flow patterns within the pore space were examined in Darcy and non-Darcy regimes: bypassing streams and steady state vortices were observed when Re was increased to about 10; unsteady flows were observed when Re reached about 20.

See more of this Session: Simulation of Complex Subsurface Processes I
See more of this Group/Topical: Topical D: Chemical Engineering in Oil and Gas Production and Other Complex Subsurface Processes

(246d) Modeling and Analysis of Wormhole Formation in Carbonates with Diverting Acids

Ram R. Ratnakar, Department of Chemical and Biomolecular Engineering, University of Houston, Houston, TX, Nitika Kalia, Halliburton, Duncan, OK and Vemuri Balakotaiah, Department of Chemical and Biomedical Engineering, University of Houston, Houston, TX

We develop a semi-empirical complete rheological model for diverting acids that accounts for the effect of change in temperature, shear rate/pressure drop, and pH/concentration. A previously developed two-scale continuum model is modified to include non-Newtonian and diverting behaviors and used with the rheological model to analyze the stimulation of single/dual core set-ups with both Newtonian and diverting acids at constant injection rate. The acidization curve is found to be shallower for diverting acids as compared to Newtonian acids. We show that an efficient stimulation of dual core set-ups with Newtonian acids can be achieved if it is operated in uniform dissolution region based on total injection rate. Finally, we show that when dual core set-ups are stimulated with diverting acids, flow is diverted from high-perm core to low-perm core and from low-perm core to high-perm core several times until the exit concentration exceeds a certain value that depends on the fluid characteristics.

See more of this Session: Simulation of Complex Subsurface Processes I
See more of this Group/Topical: Topical D: Chemical Engineering in Oil and Gas Production and Other Complex Subsurface Processes

A Coupled Control-Volume Finite Element Geomechanical Model for Discrete-Fracture Reservoir Simulators

Nan Zhao, University of Utah, Salt Lake City, UT
Milind D. Deo, University of Utah, Salt Lake City, UT
John McLennan, University of Utah, Salt Lake City, UT

Abstract

The effects of geomechanical parameters on reservoir properties are usually not considered in conventional reservoir simulation.This may have ramifications in reservoirs where compaction drive is essential or for reservoirs where fracture flow is dominant and is strongly pressure sensitive. Discrete-fracture network representations of a fractured reservoir are examples of the latter; geomechanical effects are expected to have significant impact on production performance. In this paper, we systematically explore a number of different methodologies for incorporating stress dependence into reservoir properties. A finite element, geomechanical model based on force balance principles is interfaced with a control-volume finite element, discrete-fracture reservoir simulator. The geomechanical module that has been developed is able to consider any standard thermo-poroelastic constitutive relationship or stress-strain correlation, and accounts for both fracture and matrix constitutive properties. The pressure and saturation information from the flow model are communicated to the geomechanical model and the matrix/fracture properties are updated. Several different approaches for coupling are demonstrated and compared; from explicit exchange to communication between modules. Examples from fractured tight gas formations are used to demonstrate the model and for assessing the importance and effectiveness of each type of coupling. It is shown that the relevant geomechanical information is incorporated into the reservoir model in a computationally efficient manner by allowing timely communication, not necessarily after every time step. The impact of aperture changes of the fractures is shown to be considerable - resulting in significant local changes in effective permeability and connectivity during production. The influence of fracture aperture change, associated with the geomechanical module on capillarity and other rock-fluid phenomena lead to a number of interesting observations concerning water block, fluid trapping, producible rates and ultimate production. It is significant in dealing with dynamic changes in fracture networks in fractured reservoirs.

Introduction

A finite element geomechanical model coupling with a conventional reservoir simulator, which has the capability to coupling with various reservoir models, is introuced in the Utah Finite Element simulator (UFES). The new simulator is designed to run the black oil reservoir model with geomechanics as well as thermal typed model. And it provides several choices of coupling scheme include soft coupling and hard coupling. All thes features makes UFES a simulator that can handle various of applications.

UFES Frame Work and Governing Equations

The UFES is a highly modularized simulation frame work, and it has several specific mofules with different features. The various combations of different modules enable the UFES to handle multiple applications. The major module of UFES is the reservoir module and the geomechanical module, under each module a variety of submodules is employed to handle specific problem.

Reservoir Model

The UFES maily employed two reservoir models, the black oil model[1] and the compositional k value thermal model. Both models are based on discreted frature network representation of reservoir and work with several discretization methods. The governing equation for a black oil model is:

$$\nabla \cdot (\frac{Kk_{rl}}{B_l \mu_l} \nabla \varphi_l) = \frac{\partial}{\partial t}(\phi \frac{S_l}{B_l}) + q_l \tag{1}$$

φ_l is the potenial of the species l:

$$\varphi_l = p + \gamma Z \tag{2}$$

K is the absolute preambility and k_{rl} is the relative preambility. B is the volumetric factor. ν is the viscousity.q is the source term of each species, in the reservoir simulation this term is generally refer to the injection or production well.

The governing equation for compositional model[2] is:

$$\frac{\partial (U_p + U_{rock})}{\partial t} = \nabla \cdot \bar{\bar{\mathbf{K}}}_{\mathbf{c}} \nabla T - \nabla \cdot (\sum_{p=1}^{Nf} \hat{H}_p \rho_p \vec{v}_p) + Q_e + Q_{loss} \tag{3}$$

where U_p and U_{rock} are

$$U_p = \sum_{p=1}^{Np} \phi S_p \rho_p \hat{U}_p \tag{4}$$

$$U_{rock} = (1-\phi)\bar{U}_{rock} \tag{5}$$

Geomechanical Model

The geomechnical model is based on the assumption that the equilibrium state of mechanical is enforced in the reservoir. And the equation is:

$$\nabla \cdot \sigma + \rho \vec{b} = 0 \tag{6}$$

where σ is a tensor which represents stresses (symmetric):

$$\begin{bmatrix} \sigma_{x,x} & \sigma_{x,y} & \sigma_{x,z} \\ \sigma_{y,x} & \sigma_{y,y} & \sigma_{y,z} \\ \sigma_{z,x} & \sigma_{z,y} & \sigma_{z,z} \end{bmatrix}$$

Effective stress law is applied: $\sigma = \sigma' - \alpha\delta_{i,j}P$ where σ' is the effective stress,and P can be write as the saturation weighted pore pressure[3] ($P = \sum_{i=1}^{n} s_i p_i$).

$$\nabla \cdot (\sigma' - \alpha\delta_{i,j}P) + \rho \vec{b} = 0 \tag{7}$$

The density is defined as: $\rho = (1 - \phi^*)\rho_s + \phi^*\rho_f$ Different models can be applied to the stress like linear elastic model, linear poro-elastic model and so on.In the simulation, a poroelastical model is applied:

$$\sigma_{i,j} = 2G\varepsilon_{i,j} + 2G\frac{\nu}{1-2\nu}\varepsilon_{k,k}\delta_{i,j} - \alpha p\delta_{i,j} \tag{8}$$

ε is the volumetric strain and defines as $\varepsilon_i = \frac{\partial u_i}{\partial x_i}$ and $\varepsilon_{k,k} = \sum_{i=1}^{i=3} \frac{\partial u_i}{\partial x_i}$. G is shear module, α is Biot-Willis coefficient and ν is Poission's ratio.

Coupling schemes

In general, there are two types of coupling schemes, soft coupling and hard coupling. Soft coupling means the geomechanical effect is not directly invoved in the reservoir simulation. Hard coupling method integrate geomechanical model in the reservoir simulator. Both of these two methods need some parameters to connection geomechanical model and reservoir model. One set of the parameter is porosity and permeabilty:

First the ture porosity of the reservoir is introduced[4]:

$$\phi^* = 1 - (1 - \phi_0)e^{-\varepsilon_v}$$

or in linear form:

$$\phi = (\frac{1-\phi_0}{1-\phi^*})\phi^*$$

The equation above is come from the continuity of the rock:

$$\frac{\partial \rho_s(1-\phi)}{\partial t} + \nabla((1-\phi)v_s) = 0$$

where $\nabla v_s = \nabla\frac{du_s}{dt}$, $\nabla u_s = \varepsilon_v$

The permeability can be ralated to the geomechanic factor[4]:

$$\frac{k}{k_0} = \frac{1 + \frac{\varepsilon_v}{\phi_0}}{1 + \varepsilon_v}$$

where $\varepsilon_v = \frac{\delta V}{V} = \varepsilon_{xx} + \varepsilon_{yy} + \varepsilon_{zz}$ and $\varepsilon_{ij} = \frac{1}{2}(\frac{\partial u_i}{x_j} + \frac{\partial u_j}{x_i})$ Through these information, the impact of geomechanical effect is introduced in the reservoir model.

On the other hand, the geomechanical model is related to the reservoir model by the effective stress formulation. Also the geomechanical model is assumed to be at equilibrium all the time through the reservoir life. Based on the coupling parameter, soft and hard coupling can be applied.

Soft Coupling

To take into account of the geomechanical effect with out much change of the reservoir model. The "artifical" permeability and porosity can be introduced in the soft coupling scheme. In this way, the reservoir permeability and porosity is set to be a function of pressure (may plus saturation) based on the external information of this reservoir. This external information can be some real data and some pure mechanical simulation result of reservoir.

A soft coupling scheme is developed based on Itasca 3DEC simulation result for the multiphase flow simulation In the simulator the permeability and porosity is treaed from a look-up table generated from the geomechanical information of 3DEC. The pore pressure is using as a standard to maching these two simulation.

Hard Coupling

For the hard coupling approach, the geomechanical model is integrated in the reservoir simulator. The geomechanical equation set and flow equation set can be solved either "separately" or "together" depend on the real problem.

For the geomechanical model, a general virtual work formulation can be derived:(neglect the body force for convenience, acttually it just gravity in most case which only in z dirction:

$$\int_\Omega \delta\vec{\varepsilon}^T \Delta\sigma' d\Omega - \alpha \int_\Omega \delta\vec{\varepsilon}^T \Delta p d\Omega - \int_{\Gamma_\sigma} \delta\vec{u}^T \Delta\vec{t} d\Gamma = 0$$

Use the strain stress equation:

$$(\int_\Omega B^T D B d\Omega)(\Delta\vec{u}) - \alpha(\int_\Omega B^T \vec{m} N d\Omega)(\Delta p) - \int_{\Gamma_\sigma} \delta\vec{u}^T \Delta\vec{t} d\Gamma = 0$$

The coupling parameter is the volumetric strain.

References

1. Fu, Yao (2007), "Multiphase Control Volume Finite Element Simulation of Fracured Reservoirs", PhD thesis, University of Utah

2. Huang,Chung-Kan (2009), "Development of A General Thermal Oil Reservoir Simulator Under A Modularized Framework", PhD thesis, University of Utah

3. Zienkiewicz, O. C. and Shiomi, T(1984), "Dynamic behaviour of saturated porous media; the generalized biot formulation and its numerical solution", Internation Journal for Numerical and Analytical Method in Geomechanics, 87:71-96

4. L.K. Thomas and L.Y. Chin and R.G. Pierson and J.E. Sylte(2002), "Coupled Geomechanics and Reservoir Simulation" in SPE Annual Technical Conference and Exhibition

Multiphase Compositional Space Parameterization for General-Purpose Thermal-Compositional Simulation

Alireza Iranshahr, Denis Voskov and Hamdi Tchelepi, Energy Resources Engineering, Stanford University, Stanford, CA

Introduction

Thermal and gas injection processes are among the most commonly used techniques to improve the recovery of oil. In order to obtain reliable predictions based on the numerical simulation of such processes, a thermal-compositional model is used, where the partial differential equations for the conservation of mass, energy, and momentum are solved together with the equations of thermodynamic equilibrium (Aziz and Wong, 1989).

Coupling thermodynamics with the conservation equations is a major challenge in the development of a general-purpose thermal-compositional simulation framework. The partial differential equations are highly nonlinear and tightly coupled to the phase equilibrium relations. In order to deal with thermal-compositional models of practical utility, accurate treatment of the thermodynamic phase behavior is crucial (Farouq-Ali and Abou-Kassem, 1989).

Simplified compositional (or modified black-oil) models have been developed in order to account for the thermodynamic phase behavior in reservoir flow simulation (Bolling, 1987; Whitson et al., 1988). One of the most common approaches is the so-called 'constant K-values' method, where it is assumed that the component equilibrium factors, K-values, are functions of temperature and pressure only, and therefore, thermodynamic equilibrium can be efficiently modeled using tables of K-values (see Aziz et al. (1987) for a practical example). However, the accuracy of the constant K-values approach is limited to immiscible displacements at relatively low pressure, where the assumption of composition-independent K-values is valid. In this work, we employ an Equation of State (EOS) method to represent the phase behavior associated with thermal-compositional flow simulation.

Phase equilibrium calculations can be the most expensive kernel in a compositional flow simulation. Given an overall composition at some pressure and temperature conditions, the number of phases and their compositions must be determined for every control volume (gridblock) at each time step. The conventional approach is based on: (1) performing a phase-stability test for single-phase gridblocks (Michelsen, 1982a), and (2) solving the fugacity constraints (Michelsen, 1982b) together with the coupled nonlinear flow equations for gridblocks that have more than one phase. Robust algorithms for stand-alone calculations of multiphase stability and flash procedures have been described by several authors (see Michelsen and Mollerup (2004) for instance). However, previous developments for phase behavior computations coupled with flow and transport have basically focused on methodologies that are suitable for only two-phase compositional simulation.

With the objectives of improved performance and stability of EOS-based phase behavior for two-phase compositional simulation, a tie-line based Compositional Space Parameterization (CSP) framework has been developed (Voskov and Tchelepi, 2009b). An effective way of utilizing the concept of CSP for general-purpose simulation is the Compositional Space Adaptive Tabulation (CSAT) framework in which tie-line parameterization is used to accelerate the convergence of EOS-based phase-behavior calculations (Voskov and Tchelepi, 2009a). CSAT is based on theoretical findings from Method of Characteristics (MoC) solutions of gas injection problems under ideal assumptions (Franklin M. Orr, 2007; Entov, 1997). The MoC solutions show that only a few (key) tie-lines determine the structure of the solution route in compositional space. The MoC-based theory has been recently extended to thermal (Zhu, 2003) and three-phase (LaForce et al., 2008b) displacements.

CSAT adaptively parameterizes the compositional space around a given composition (i.e., calculates the tie-lines through the composition at discrete pressure levels). The tie-line tables are then used to identify the phase-state of compositions encountered in the course of the simulation. Linear interpolation in pressure is performed to obtain a tie-line for a composition with the specified pressure value (Voskov and Tchelepi, 2009b).

The initial development of CSAT for immiscible displacements has been generalized for multi-contact miscible problems (Voskov and Tchelepi, 2009a). It is shown that adaptive tabulation of the critical tie-lines significantly improves the efficiency and accuracy of multi-contact miscible simulation. Generalization of CSAT to thermal problems, where tabulation and interpolation as a function of temperature and pressure is performed, has been presented in Iranshahr et al. (2010). Extension to multiphase systems is described in Voskov and Tchelepi (2009c). The wide range of the numerical examples presented in these references shows that CSAT is an effective method for phase behavior modeling in general-purpose compositional simulation.

In this paper, we present an EoS-based negative-flash method for systems that can form an arbitrary number of phases. The proposed algorithm is guaranteed to converge, and it is integrated with our multiphase CSAT framework for phase behavior calculations. We also describe the continuity of compositional space parameterization using tie-simplexes. Based on these ideas, the CSAT framework is able to capture the complex multiphase behavior in the vicinity of a given composition at different temperature and pressure conditions. We also describe a generalization of the CSAT framework for three-phase systems, and we demonstrate accurate phase-state identification in complex multicomponent systems.

EoS Based Multiphase Negative-flash

We start with the formulation of the multiphase negative-flash framework based on an Equation of State (EoS). Let N_p represent the maximum number of the phases that can coexist for a given system at the specified temperature and pressure. Here, we consider the partitioning of mixture made up of N_c components into N_p phases at equilibrium. The equations can be written as

$$\hat{f}_{i,0} - \hat{f}_{i,j} = 0, \quad i = 0, \ldots, N_c - 1, \; j = 1, \ldots, N_p - 1, \tag{1}$$

$$z_i - \sum_{j=0}^{N_p-1} x_{i,j}\nu_j = 0, \quad i = 0, \ldots, N_c - 1, \tag{2}$$

$$\sum_{i=0}^{N_c-1} (x_{i,0} - x_{i,j}) = 0, \quad j = 1, \ldots, N_p - 1, \tag{3}$$

$$1 - \sum_{j=0}^{N_p-1} \nu_j = 0, \tag{4}$$

where z_i denotes the overall composition of component i. The fugacities in Eqs. (1), $\hat{f}_{i,j}$, are computed using an EoS model (Peng and Robinson, 1976). The above $N_p(N_c + 1)$ equations are solved for the fraction (ν_j) and composition ($x_{i,j}$) of each phase. The pure phases define the vertices of a tie-simplex in the compositional space. Noting that the actual number of phases at equilibrium may be smaller than N_p, the phase fractions are not restricted to $[0, 1]$.

Newton's method can be used to solve Eqs. (1-4). Convergence of the Newton iterations depends on the initial guess. A good initial guess is obtained using Successive Substitution Iterations (SSI) (Whitson and Michelsen, 1989). During each iteration of SSI, component fugacities are held fixed, and Eqs. (2-4) are solved simultaneously. Here, we present a generalization of the two-phase negative-flash approach (Whitson and Michelsen, 1989). The overall mole fraction of a component, z_i, is given by:

$$z_i = x_{i,0}\nu_0 + \sum_{j=1}^{N_p-1} x_{i,j}\nu_j, \qquad i = 0, \ldots, N_c - 1, \tag{5}$$

where we have arbitrarily chosen a phase as the 'base' phase, and we denoted its index by 0. Using

$$\nu_0 = 1 - \sum_{j=1}^{N_p-1} \nu_j, \tag{6}$$

$$x_{i,j} = K_{i,j}x_{i,0}, \qquad i = 0, \ldots, N_c - 1; \quad j = 1, \ldots, N_p - 1, \tag{7}$$

one obtains

$$z_i = x_{i,0}\left(1 + \sum_{j=1}^{N_p-1} \nu_j(K_{i,j} - 1)\right) = x_{i,0}m_i(\boldsymbol{\nu}), \qquad i = 0, \ldots, N_c - 1. \tag{8}$$

In order to determine the equilibrium partitioning of all components among all phases, one must solve the following coupled $N_p - 1$ equations

$$\begin{aligned} f_j(\boldsymbol{\nu}) &= \sum_{i=0}^{N_c-1} (x_{i,0} - x_{i,j}) \\ &= \sum_{i=0}^{N_c-1} \frac{z_i(1 - K_{i,j})}{m_i(\boldsymbol{\nu})} = 0, \qquad j = 1, \ldots, N_p - 1, \end{aligned} \tag{9}$$

for $N_p - 1$ unknowns, $\boldsymbol{\nu}(\equiv \{\nu_j\},\ j = 1, \ldots, N_p - 1)$.

Next, we define the domain of the full problem. For this purpose, let us assume that we have an admissible set of K-values, so that the domain of Eqs. (9) is bounded. From Eqs. (8), we conclude that $m_i(\boldsymbol{\nu})$ must be strictly positive for each component (denoted by index i). The set of 'physical' phase fractions for this problem can be represented by

$$\Omega = \{\boldsymbol{\nu} \in \mathbb{R}^{N_p-1} \mid 0 < m_i(\boldsymbol{\nu}),\ i = 0, \ldots, N_c - 1\}. \tag{10}$$

Given an admissible set of K-values, Ω is a full-dimensional bounded convex polytope in $\mathbb{R}^{N_p-1}$ (Brondsted, 1983). Next, let us consider a subproblem of the full negative-flash problem. Suppose that for an N_p-phase negative-flash problem, we want to solve for the first M_p phase fractions given that the last $N_p - M_p - 1$ phase fractions are already known:

$$\min \nu_{j,\Omega} < \nu_j < \max \nu_{j,\Omega}, \qquad j = M_p + 1, \ldots, N_p - 1, \tag{11}$$

where ν_j is a value in the range of the j-th phase fraction for the full-dimensional Ω. The equations to be solved simultaneously for $\boldsymbol{\nu}(\equiv \{\nu_j\},\ j = 1, \ldots, M_p)$ are:

$$f_j(\boldsymbol{\nu}) = \sum_{i=0}^{N_c-1} \frac{z_i(1 - K_{i,j})}{m_i(\boldsymbol{\nu}) + n_i} = 0, \qquad j = 1, \ldots, M_p, \tag{12}$$

where n_i is a fixed number given by

$$n_i = \sum_{j=M_p+1}^{N_p-1} \nu_j (K_{i,j} - 1), \qquad i = 0, \ldots, N_c - 1. \tag{13}$$

Note that $m_i(\boldsymbol{\nu})$ is a function of the M_p phase fractions only. Let

$$\Omega^{M_p} = \{\boldsymbol{\nu} \in \mathbb{R}^{M_p} \mid 0 < m_i(\boldsymbol{\nu}) + n_i,\ i = 0, \ldots, N_c - 1\}, \tag{14}$$

represent the set of physical phase fractions for this subproblem. It can be shown that a solution for the negative-flash subproblem exists, and that it is unique (Iranshahr et al., 2009a). Moreover, the subproblem has all the properties of a full negative-flash problem, so that a recursive procedure can be applied to solve the multiphase negative-flash problem.

Implementation of the Multiphase Negative-Flash

For three-phase systems, Ω is a convex polygon, and we solve for two phase fractions (ν_1, ν_2). Here, we describe a bisection based algorithm. In order to improve the speed of convergence, the bisection strategy can be combined with a gradient based scheme. Consider the following algorithm:
Step 1: Given the set of K-values, find the boundaries of the physical plane, which yield $\nu_{2,min}$ and $\nu_{2,max}$.
Step 2: Set $\nu_{2,mid} = (\nu_{2,min} + \nu_{2,max})/2$.
Step 3: Solve $f_1(\nu_1, \nu_{2,min}) = 0$, and $f_1(\nu_1, \nu_{2,mid}) = 0$ for $\nu_{1,min}$, and $\nu_{1,mid}$, respectively using bisection.
Step 4: If $f_2(\nu_{1,min}, \nu_{2,min}) \times f_2(\nu_{1,mid}, \nu_{2,mid}) > 0$, $\nu_{2,min} = \nu_{2,mid}$; else $\nu_{2,max} = \nu_{2,mid}$.
Step 5: Stop if $\nu_{2,max} - \nu_{2,min} < \varepsilon$; otherwise, go to Step 2.
Figure (1) illustrates the first four iterations of the procedure for a synthetic 15-component system (with arbitrary K-values). The SSI method for solving two-phase negative-flash problems can be generalized to three-phase systems by using this algorithm as a constant K-value flash solver at each iteration. For simplification in the following sections, we will describe the developments for multicomponent systems that can form a maximum of three phases at equilibrium.

Multi-Stage Three-Phase Negative-Flash

The three-phase negative-flash introduced in the previous section is not sufficient for determining the phase-state of a given composition. This is because the two-phase behavior outside the three-phase region may be different from the results obtained from a three-phase negative-flash (see Figure 2). Based on a three-phase negative-flash only, a mixture can be identified as stable in the single-phase region; however, its Gibbs free energy might decrease by splitting into two mixtures. To avoid such problems, a second stage using two-phase negative-flash can be employed.

In order to determine the number and compositions of the existing phases for a given overall composition at specified temperature and pressure conditions, we use a multi-stage negative-flash algorithm. In the first stage, a three-phase negative-flash is performed. If the composition is inside the tie-triangle (three-phase), its state is confirmed. Otherwise, based on the results of the three-phase negative-flash, additional two-phase negative-flash computations are necessary. Here, we assume that the tie-lines are in the plane of the tie-triangle, even though real multicomponent systems show slight deviations from this

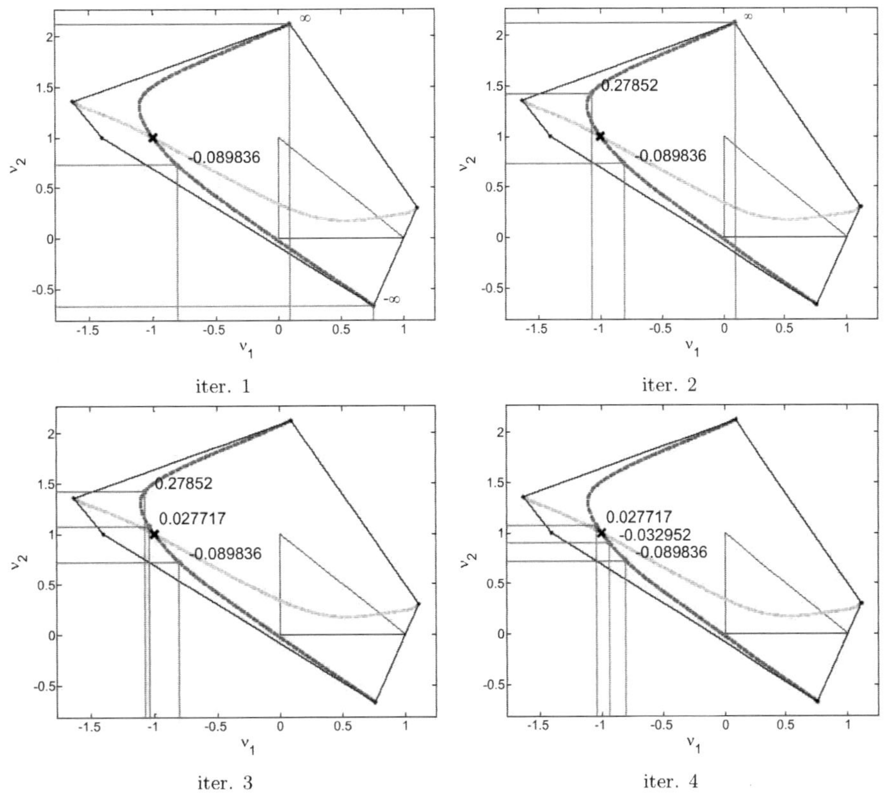

Figure 1 *The first four iterations of the methodology for a 15-component system.* Ω *is represented by the black polygon, and the blue triangle encloses three-phase mixtures.* f_1 *and* f_2 *are zero along the red and green curves, respectively. The numbers represent the values of* $f_2(\nu_1, \nu_2)$ *at* $\nu_{2,min}$, $\nu_{2,mid}$, *and* $\nu_{2,max}$ *where* $f_1(\nu_1, \nu_2) = 0$. *Note that initially the value of* f_2 *is indefinite at* $\nu_{2,min}$ *and* $\nu_{2,max}$. *These lower and upper bounds become closer as the iterations proceed.*

behavior. The compositions of the tie-triangle edges (two-phase systems) provide good initial guesses for the various two-phase negative-flashes. This is because tie-line compositions change continuously around a tie-triangle edge (Iranshahr et al., 2009b).

The three-phase negative-flash procedure can be used to generate phase diagrams (see Figure 3), and this is demonstrated using two examples: (1) a sour gas example, and (2) a Maljamar reservoir oil (Haugen et al., 2007). The accuracy of our numerical results is verified by a commercial simulator CMG (2009). We consider dilution of both mixtures with pure CO_2. That is, for every mixture which results from the addition of CO_2 to the initial feed, we increase the pressure and record the three-phase behavior. The pressure is increased until the parameterizing tie-triangle degenerates to a tie-line. Figure (3) shows the locus of the three-phase boundary, as well as, the tie-triangle degeneration locus. We note that no phase-stability test is required, since the three-phase negative-flash is capable of determining the bounds of the three-phase region accurately.

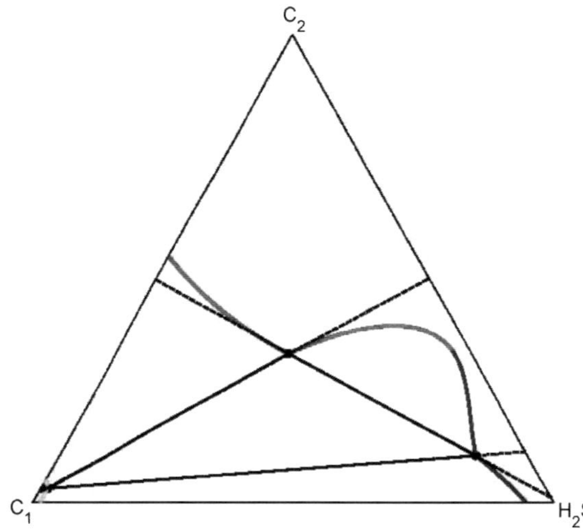

Figure 2 *Complete parameterization of* $\{C_1, C_2, H_2S\}$ *at* $T = 180\,\mathrm{K}, p = 15$ *bar. The accurate two-phase behavior is shown by the solid curves outside the three-phase region.*

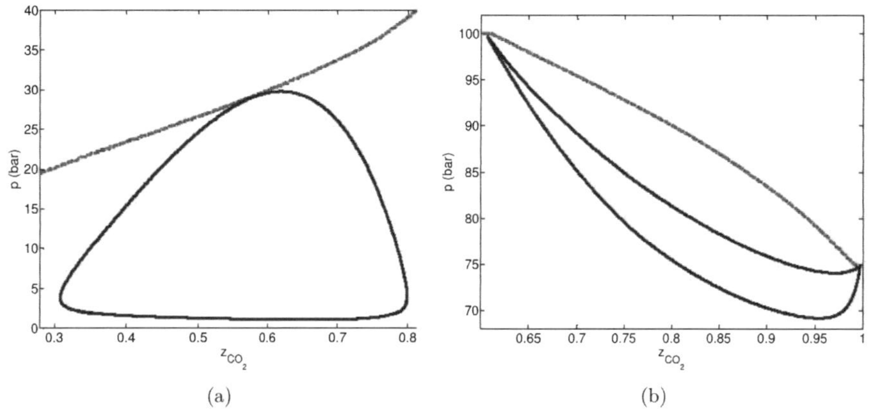

Figure 3 *Phase diagram for the three-phase region (solid), and the locus of tie-triangle degeneration (dashed): (a) Sour gas example,* $T = 178.8\,\mathrm{K}$*, (b) Maljamar reservoir oil,* $T = 305.4\,\mathrm{K}$.

Continuity of Compositional Space Parameterization

Here, we describe the continuity of parameterization using tie-simplexes. We can show that tie-simplex parameterization in compositional space is a continuous function of pressure, temperature and composition (Iranshahr et al., 2009b).

Continuation of the tie-simplex space justifies a tie-simplex based approach for compositional flow modeling, where interpolation in pressure and temperature is performed during a simulation run using a limited number of tie-simplexes. Figure (4) describes the parameterization continuity for two-phase systems. In the first example (see Figure 4-a), we plot an isothermal table of tie-lines ($T = 375$ K) for a fixed composition. The mixture is initially in the two-phase region. With increase in pressure, the composition becomes single-phase, and the parameterizing tie-line eventually degenerates at $p = 145\ bar$. We also show how parameterization of a four-component mixture changes with temperature at fixed pressure (see Figure 4-b). For this example, the composition is initially single-phase and enters the two-phase region with increase in temperature. The parameterizing tie-line becomes critical at $T = 550$ K.

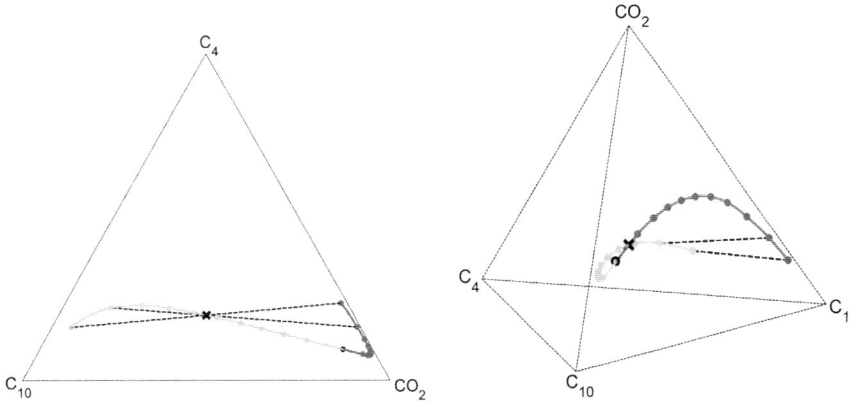

(a) $T = 375$ K, $p = (10, 145)\ bar$ (b) $p = 100\ bar$, $T = (250, 550)$ K

Figure 4 *Continuity of compositional space parameterization for two-phase systems*

Three-Phase Compositional Space Parameterization

We start with an overview of the two-phase CSAT framework (Voskov and Tchelepi, 2009b,a). Given the overall composition of a gridblock, all the tie-line tables are checked (after interpolation in temperature and pressure) to determine if there is any tie-line that parameterizes the composition. The phase-state of the mixture is determined if its composition and the interpolated tie-line are close enough. Otherwise, a new tie-line table is parameterized and added to the list of existing tables. For multi-contact miscible displacements, critical tie-lines (at different pressure values) that parameterize the solution path are adaptively calculated and stored. An initial check in the table of critical tie-lines determines if the composition belongs to the super-critical or sub-critical regions.

For two-phase systems, a single tie-line parameterizes the two-phase region around a given overall composition. However, for three-phase systems, the entire plane defined by a tie-triangle needs to be parameterized. Note that only a few tie-triangle planes parameterize the solution path in a three-phase compositional displacement (LaForce et al., 2008a; Voskov and Tchelepi, 2009c). For a given overall composition, we calculate a tie-triangle (Δ^2), and parameterize three curves: Γ_1^2, Γ_2^2, and Γ_3^2. In general, Γ_f^n parameterizes the n-phase region on the f-th side of a tie-simplex. Each one of the Γ^2 curves parameterizes the two-phase behavior on one of the tie-triangle sides.

Here, we demonstrate the construction of a general three-phase thermal-compositional table for a given composition (for parameterization of two-phase systems, see Iranshahr et al. (2010)). We assume that the starting pressure at each temperature level is below the minimal degeneration pressure for the three-phase region. For the smallest pressure, the tie-triangle that parameterizes the given composition is calculated and stored. Starting from each edge of the tie-triangle, we parameterize the two-phase region by solving for each Γ_f^2.

The three-phase region degenerates with increase in pressure. If the pressure is large enough, we solve for the minimal degeneration pressure of the tie-triangle that parameterizes the given composition. An iterative procedure is used to find a pressure value such that the area of the tie-triangle, A, satisfies:

$$\varepsilon_{min} < A < \varepsilon_{max}, \tag{15}$$

for some small values of ε_{min} and ε_{max}. The plane of the degenerate tie-triangle is then parameterized at higher pressure values. We note that immediately after the degeneration of the tie-triangle, a two-phase region disappears from the plane, or the length of a side of the tie-triangle becomes infinitesimal. However, all the parameterization techniques developed for sub-critical pressures are applicable for higher pressure values.

Tie-lines on a tie-triangle side may also degenerate. For such systems, parameterization of Γ_f^2 is started from the non-degenerate end. Then, we determine how far the parameterization of the tie-lines along Γ_f^2 should proceed until the length of the critical tie-line, ℓ, satisfies:

$$\varepsilon_{min} < \ell < \varepsilon_{max}, \tag{16}$$

for some small values of ε_{min} and ε_{max}. The calculated critical tie-line marks the boundary between extendible and non-extendible regions on the f-th tie-triangle side.

Phase-State Identification in Parameterized Compositional Space

Based on theoretical findings from the Method of Characteristics, only a few tables of tie-simplexes are required for a single simulation run (LaForce et al., 2008a; Voskov and Tchelepi, 2009c). In the second stage of the CSAT framework, we use tabulated tie-simplexes to identify the phase-state of mixtures that require phase stability testing (Voskov and Tchelepi, 2009a).

In the following developments, we assume that a side tie-line lies on the extension of a tie-triangle. The systems that we have considered in our studies exhibit this behavior. We note that CSAT still provides a very good initial guess for conventional phase-behavior calculations in systems that violate this assumption.

The tables of tie-simplexes are used to interpolate in temperature and pressure, then $\mathbf{z}$ is projected onto the closest tie-simplex (Voskov and Tchelepi, 2009c). In the general case, $\mathbf{z}$ is in M_p-phase state, if it is inside a Δ^{M_p-1}. Moreover, $\mathbf{z}$ is single-phase, if it is outside all tie-simplexes. These state identification rules are based on the fact that an M_p-phase negative flash can only determine the state of a mixture in the M_p-phase region. Next, we describe state identification procedures in degenerate steam systems. Consider a three-component steam example at $T = 550\,\mathrm{K}$ (see Figure 5-a). The three-phase region becomes critical at $p = 110\,bar$. With further increase in pressure, the gas-oil region shrinks until it disappears completely from the plane at $p = 148.9\,bar$.

From the continuity of parameterization in degenerate space, the critical gas-oil tie-lines change continuously with pressure. We calculate and store the critical tie-lines at discrete pressure values between the minimal three-phase degeneration pressure and the gas-oil

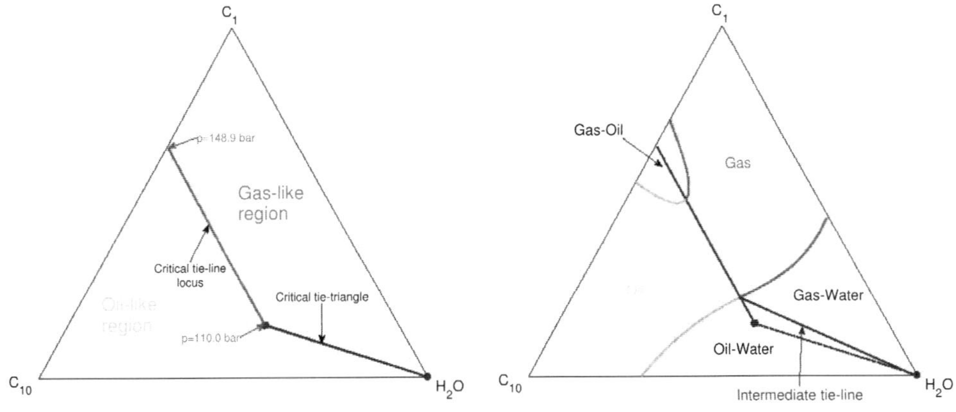

(a) (b)

Figure 5 *(a) Critical locus for* $\{C_1, C_{10}, H_2O\}$ *at* $T = 550\,\mathrm{K}$*; the locus of critical oil-gas tie-lines is extended from* $p = 110\,bar$ *to* $p = 148.9\,bar$*. (b) Phase diagrams at* $p = 137\,bar$

disappearance pressure. Note that the locus of the critical tie-lines is almost linear on the plane of the tie-triangle. The discretized locus of critical tie-lines as well as the critical tie-triangle is then used to identify if a single-phase mixture is gas- or oil-like. This technique is based on the assumption that the phase-state of a composition on the extension of a critical tie-line does not change with an increase in pressure (Voskov and Tchelepi, 2009a). We use this critical boundary to separate water-oil and water-gas tie-lines with an intermediate tie-line, which connects the water phase to a point on the critical tie-line locus (see Figure 5-b). If the water tie-lines shrink in size, the critical tie-triangle serves as the intermediate tie-line.

CSAT Examples

We show that the generalized CSAT framework accurately captures degeneration of the multiphase regions across a wide range of pressure. In an isothermal three-phase table, only two sets of tie-lines parameterize the plane of a degenerate tie-triangle. With further increase in pressure, a two-phase region disappears from the plane. We use an iterative procedure to calculate the pressure at which the two-phase region disappears. That is, we find a pressure value such that the disappearing two-phase region is represented by a trivial tie-line (the length of which satisfies Eq. 16). Figure 6 describes the general structure of an isothermal three-phase table. Note that once the multiphase region is represented by only one set of tie-lines, we use the standard two-phase CSAT approach for phase-state identification (Voskov and Tchelepi, 2009b).Figure (7) demonstrates the accuracy of CSAT close to the pressure at which the three-phase region disappears. Our computations are compared with results from a reliable commercial software CMG (2009).

Parameterization of four-phase systems is significantly more complicated than three-phase problems. We first calculate the tie-tetrahedron that parameterizes a given composition. Four curves, Γ^3, parameterize the three-phase region around the tie-tetrahedron. Moreover, different two-phase regions in the system are calculated using six surfaces, Γ^2. This way, all the multiphase regions on the extension of the tie-tetrahedron are parameterized. Figure (8) shows different sets of tie-triangles around a tie-tetrahedron.

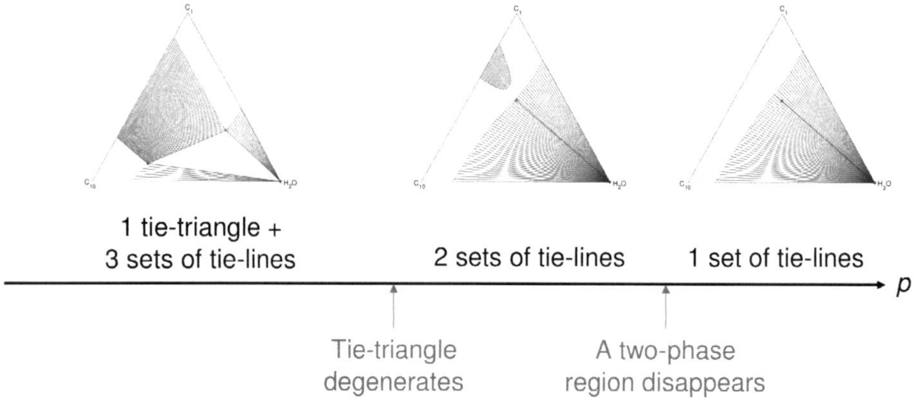

Figure 6 *General structure of an isothermal three-phase table. The three-component systems are calculated at* $T = 520\,\mathrm{K}$ *and* $p = 75, 180, 250\,bar$*, respectively.*

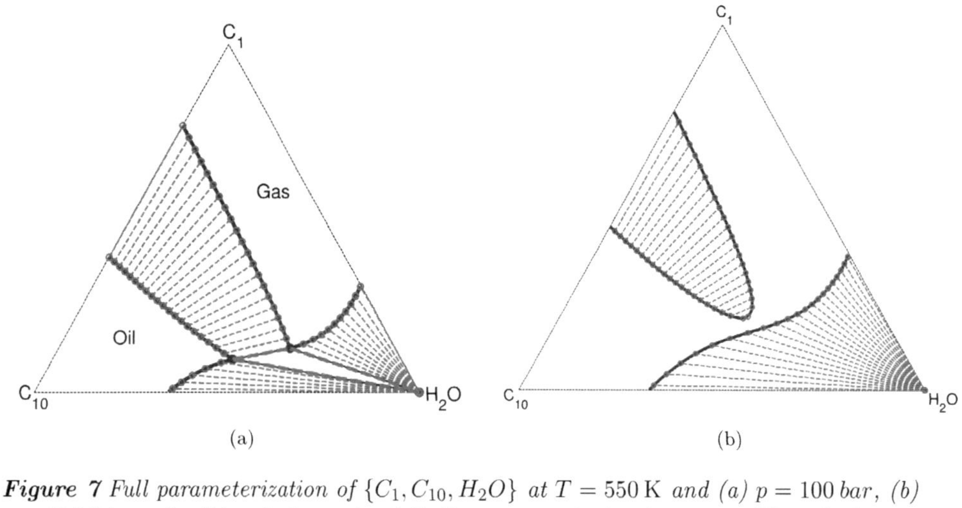

Figure 7 *Full parameterization of* $\{C_1, C_{10}, H_2O\}$ *at* $T = 550\,\mathrm{K}$ *and (a)* $p = 100\,bar$*, (b)* $p = 113.8\,bar$*. In (b) only two sets of tie-lines parameterize the space. The calculations are performed using the PR-EoS, tie-triangle degenerates at* $p = 110.0\,bar$ *and the oil-gas region disappears at* $p = 148.9\,bar$*. Compare the phase diagrams in blue with the standard calculations (red tie-lines)*

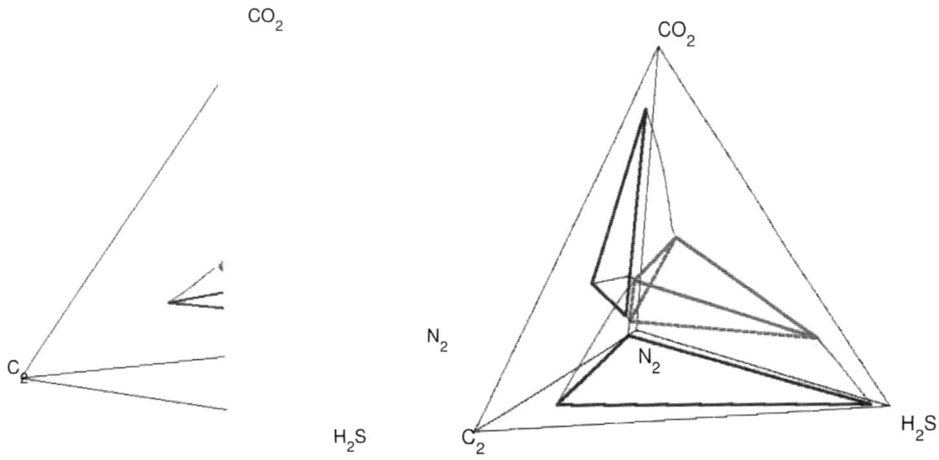

Figure 8 *Four-phase compositional space parameterization at* $T = 140\,\mathrm{K}$ *and* $p = 200\,bar$. *Each figure shows the tie-tetrahedron and two sets of side tie-triangles.*

Next, we demonstrate accurate phase-state identification for several steam systems. Consider the three-component steam example in Figure (7-a). We choose two compositions and run CSAT to identify the state of the compositions along the dilution line (see Figure 9-a). For the first composition, CSAT constructs a three-phase table as described earlier. The phase-state of all the compositions are then determined using this table. Note that a single table parameterizes the entire three-component system. A two- or three-phase composition is parameterized by a tie-line, or a tie-triangle, respectively. A single-phase composition is outside all tie-lines and the tie-triangle. Additional two- or three-phase negative-flash calculations may be required in order to identify the phase-state of mixtures that are too close to the phase boundaries accurately (Voskov and Tchelepi, 2009a).

With increase in pressure, the three-phase region degenerates, and the gas-oil region exhibits critical behavior (see Figure 9-b). Here, we also use CSAT to determine the phase state of a dilution line. Note that the locus of the critical tie-lines, which are at different pressures, separates single-phase gas and oil compositions (see Figure 5). Next, we consider a four-component steam system (see Figure 10-a for the full parameterization). We choose two compositions in this system and run CSAT along the dilution line. As shown in the figure, only a few parameterized planes are required for phase-state identification.

Conclusions

We formulated the multiphase EOS-based negative-flash framework. The Successive Substitution Iterations (SSI) and Newton methods for solving the composition-dependant phase-splitting problem can be generalized for systems with an arbitrary number of phases. Noting that a unique solution of the negative-flash procedure exists for each iteration of SSI, we employ a robust solution framework that is guaranteed to converge. We presented several challenging multiphase multi-component systems to demonstrate the effectiveness of the generalized negative-flash method. Our results indicate that the N_p-phase negative-flash can be used to identify accurately the phase-state of mixtures that partition into N_p equilibrium phases. For mixtures that partition into a lower number of phases at given

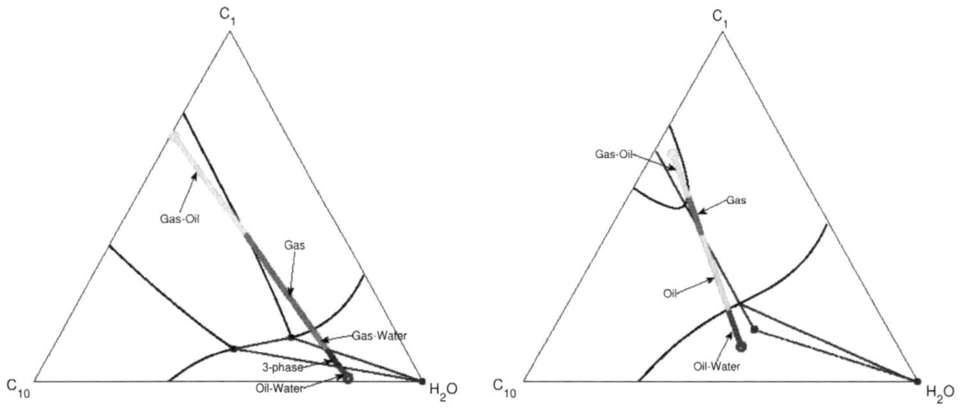

(a) (b)

Figure 9 *Phase state identification at* $T = 550$ K *for the dilution line between (a)* $\{C_1(0.7), C_{10}(0.29), H_2O(0.01)\}$ *and* $\{C_1(0.01), C_{10}(0.19), H_2O(0.8)\}$ *at* $p = 100\,bar$, *and (b)* $\{C_1(0.65), C_{10}(0.3), H_2O(0.05)\}$ *and* $\{C_1(0.1), C_{10}(0.4), H_2O(0.5)\}$ *at* $p = 137\,bar$.

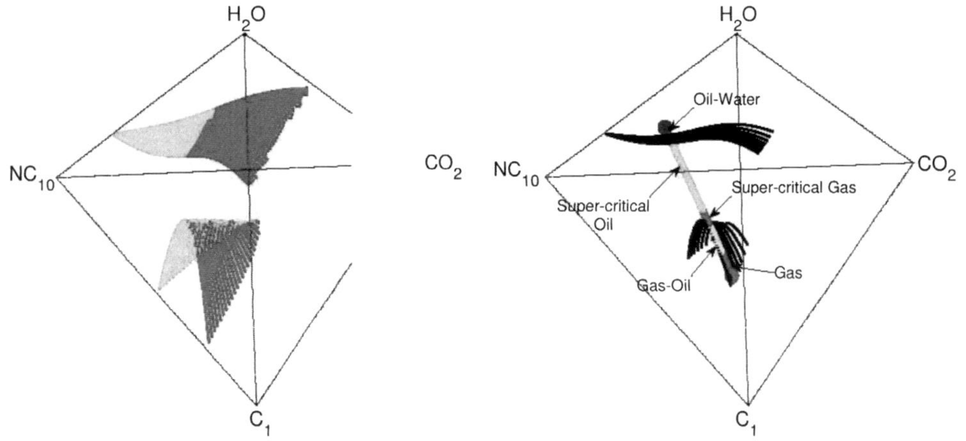

(a) (b)

Figure 10 *(a) Full parameterization of* $\{C_1, CO_2, C_{10}, H_2O\}$ *at* $T = 550$ K *and* $p = 137\,bar$, *(b) Phase state identification for the dilution line between* $\{C_1(0.05), C_{10}(0.45), CO_2(0.42), H_2O(0.08)\}$ *and* $\{C_1(0.47), C_{10}(0.265), CO_2(0.04), H_2O(0.225)\}$.

pressure and temperature conditions, we proposed a multi-stage negative-flash procedure.

We also described the theory of parameterization continuity for the multiphase Compositional Space Adaptive Tabulation (CSAT) framework. We extended the CSAT framework for thermal-compositional systems that can form an arbitrary number of phases. We presented the details of three-phase compositional space parameterization (e.g., multiphase degeneration, two-phase disappearance) over wide ranges of pressure. We described phase-state identification procedures in degenerate steam systems. Our numerical examples indicate that CSAT is as reliable as conventional procedures for phase-state identification in multiphase systems.

Acknowledgements

Thanks to the SUPRI-B Industrial Affiliates program on reservoir simulation at Stanford University for financial support.

References

Aziz, K., Ramesh, A. and Woo, P. [1987] Fourth spe comparative solution project: Comparison of steam injection simulators. *Journal of Petroleum Technology*, **39**, 1576–1584.
Aziz, K. and Wong, T. [1989] Considerations in the development of multipurpose reservoir simulation models. *proceedings of the 1st and 2nd International Forum on Reservoir Simulation, Alpbach, Austria.*
Bolling, J. [1987] Development and application of a limited-compositional, miscible flood reservoir simulator. *SPE Symposium on Reservoir Simulation, 1-4 February 1987, San Antonio, Texas.*
Brondsted, A. [1983] *An Introduction to Convex Polytopes.* Springer-Verlag New York Inc.
CMG [2009] *WinProp users's guide.*
Entov, V. [1997] Nonlinear waves in physicochemical hydrodynamics of enhanced oil recovery. *Multicomponent flows, Proc. of the International Conference Porous Media: Physics, Models, Simulations, Moscow, Russia.*
Farouq-Ali, S. and Abou-Kassem, J. [1989] Simulation of thermal recovery processes. *proceedings of the 1st and 2nd International Forum on Reservoir Simulation, Alpbach, Austria.*
Franklin M. Orr, J. [2007] *Theory of Gas Injection Processes.* Tie-Line Publications.
Haugen, K., Sun, L. and Firoozabadi, A. [2007] Three-phase equilibrium calculations for compositional simulation, spe 106045. *proceedings SPE Reservoir Simulation Symposium, Woodlands, TX.*
Iranshahr, A., Voskov, D. and Tchelepi, H.A. [2009a] A generalized negative flash procedure for phase equilibrium computations related to carbon dioxide sequestration. *proceedings AIChE Annual Meeting, Nashville, TN.*
Iranshahr, A., Voskov, D. and Tchelepi, H.A. [2009b] Theory and practice of tie-simplex parameterization for compositional reservoir simulation. *proceedings AIChE Annual Meeting, Nashville, TN.*
Iranshahr, A., Voskov, D. and Tchelepi, H.A. [2010] Tie-simplex parameterization for eos-based thermal compositional simulation. *SPE Journal*, **15**, 537–548.
LaForce, T., Jessen, K. and Orr, F. [2008a] Analytical solutions for four-component gas/water/oil displacements: Part ii. example solutions. *Transport in Porous Media*, **72**, 83–96.
LaForce, T., Jessen, K. and Orr, F. [2008b] Four-component gas/water/oil displacements in one dimension: Part i. structure of the conservation law. *Transport in Porous Media*, **71**, 199–216.
Michelsen, M. [1982a] The isothermal flash problem: Part i. stability. **9**, 1–19.
Michelsen, M. [1982b] The isothermal flash problem: Part ii. phase-split calculation. **9**, 21–40.
Michelsen, M. and Mollerup, J. [2004] *Thermodynamic Models: Fundamentals & Computational Aspects.* Tie-Line Publications.
Peng, D. and Robinson, D. [1976] A new two-constant equation of state. *Ind. Eng. Chem. Fundam.*, **15**, 59–64.
Voskov, D.V. and Tchelepi, H.A. [2009a] Compositional space parameterization: Multicontact miscible displacements and extension to multiple phases. *SPE Journal*, **14**, 441–449.
Voskov, D.V. and Tchelepi, H.A. [2009b] Compositional space parameterization: Theory and application for immiscible displacements. *SPE Journal*, **14**, 431–440.
Voskov, D.V. and Tchelepi, H.A. [2009c] Tie-simplex based mathematical framework for thermodynamical equilibrium computation of mixtures with an arbitrary number of phases. *Fluid Phase Equilibria*, **283**(1-2), 1 – 11.
Whitson, C., da Silva, F. and Soreide, I. [1988] Simplified compositional formulation for modified

black-oil simulators. *proceedings of the SPE Annual Technical Conference and Exhibition, SPE 18315.*
Whitson, C. and Michelsen, M. [1989] The negative flash. *Fluid Phase Equilibria*, **53**, 51–71.
Zhu, J. [2003] *Multicomponent Multiphase Flow in Porous Mdedia with Temperature Variation or Adsorption*. PhD Thesis, Stanford University.

Accurate Representation of EoS Related Non-Linearities for Compositional Flow Simulation

Rustem Zaydullin, Denis Voskov and Hamdi Tchelepi, Energy Resources Engineering, Stanford University, Stanford, CA

Introduction

Compositional simulation is necessary for modeling Enhanced Oil Recovery (EOR) processes. To simulate compositional processes we need to solve a coupled system of nonlinear conservation laws for multi-component multiphase flow and transport with a nonlinear system of thermodynamical constraints that represents the phase behavior of the mixture. An Equation of State (EoS) model is usually employed to describe the phase behavior (Aziz and Wong, 1989; Lake, 1989). For a given temperature, pressure, and overall composition, EoS computations define the phase state (stability test) and composition of each phase (flash calculation). Since EoS computations must be performed for each computational cell and for each global Newton iteration, they can consume a significant fraction of the total computational cost.

There are several ways to improve the efficiency of compositional flow simulation. In cases where the phase behavior is a weak function of composition, a model based on constant K-values can be used, in which the K-values depend on pressure only (Bolling, 1987). Another possibility is to limit the number of components and represent the phase behavior using simplified relations (Todd and Longstaff, 1972; Pope and Nelson, 1972).

A general approach for phase-behavior representation of the two-phase compositional systems of practical interest was suggested by Entov et al. (2001). Their method is based on projection of the original compositional problem into tie-line space, where simple polynomial approximations of the tie-line space are used to capture the important features of the two-phase region (Voskov, 2002). Voskov and Tchelepi (2009) used linear interpolation in tie-line space, which improved the accuracy and versatility of the approach. They refer to the approach as CSP (Compositional Space Pramaterization) based flow simulation.

The CSP approach is based on parameterizing the entire compositional space. It is known that gas-injection displacement processes involve a limited number of (key) tie-lines (Franklin M. Orr, 2007). Moreover, it was shown that for purely hyperbolic 1D problems, the projection of the solution into tie-line space is invariant with respect to the hydrodynamic parameters of the problem (Entov, 1997; Voskov and Entov, 2001). Later it was demonstrated that for compressible, multi-dimensional displacements in porous media, projection of the solution in tie-line space is nearly invariant, even though a larger number of tie-lines may be involved due to the numerical dispersion associated with the computed solutions.

In Voskov and Tchelepi (2009b, 2008), the fact that only a few tie-lines are needed was used to improve phase behavior computations for general purpose compositional simulation. The Compositional Space Adaptive Tabulation (CSAT) approach adaptively stores discrete tie-lines during a simulation. The collected tie-lines are used to precondition standard EoS behavior computations. However, since CSAT uses a discrete representation of only part of the compositional space, it cannot replace the EoS computations completely.

Here we present an adaptive version of the full parameterization (CSP) approach. During a simulation, we adaptively compute several supporting tie-lines and use adaptive triangulation of the tie-line space. The governing equations are cast in terms of the tie-line parameters and solved numerically. In order to achieve this, several modifications of the standard compositional framework were made, including the flash calculations. Linear interpolation based on tie-line parameters replaces EoS-based computations and leads to significant gains in computational efficiency. Based on this approach, it is also possible to compute the eigenvectors and eigenvalues that describe the displacement behavior, which can be helpful in constructing improved linear and nonlinear solvers. The applicability of this CSP method is demonstrated for several challenging compositional cases of practical interest.

Conventional Compositional Simulation Approach

The system of equations for a two-phase isothermal compositional problem can be written as:

$$\frac{\partial}{\partial t}\left(\phi\sum_{j=1}^{2}x_{i,j}\rho_j S_j\right)-\operatorname{div}\sum_{j=1}^{2}x_{i,j}\rho_j\Big(\mathbf{k}\frac{k_{rj}}{\mu_j}\nabla p\Big)+\sum_{j=1}^{2}x_{i,j}\rho_j q_j \;=\; 0,\, i=1,\ldots,n_c. \quad (1)$$

Here $x_{i,j}$ is the concentration of component i in phase j, S_j is the saturation of phase j, ρ_j – is the phase molar density, $\boldsymbol{k}$ is a permeability tensor, k_{rj} is the relative permeability, μ_j is phase viscosity, and p is pressure.

Instantaneous thermodynamic equilibrium is the usual assumption for compositional flow simulation. Normally, there are two possibilities to deal with the thermodynamic constraints: 1) solve the mass conservation equations and the thermodynamic equilibrium relation simultaneously, 2) couple the thermodynamic relations with the mass conservation equations sequentially. These approaches have their own advantages and drawbacks (Voskov et al. (2009)). Here, we follow the second approach.

The overall composition, $\mathbf{z}$, can be written as

$$z_i = x_{i,o}L + x_{i,g}(1-L), \quad i=1,\ldots,n_c, \quad (2)$$

$$L = S_o\rho_o/(S_o\rho_o + S_g\rho_g), \quad (3)$$

where L is the liquid fraction. The accumulation term of Eq. (1) can be written as

$$\phi\sum_j x_{i,j}\rho_j S_j = \phi\rho z_i, \quad (4)$$

where $\rho = S_o\rho_o + S_g\rho_g$. We solve system of Eq. (1) using p and $z_i,\ i = 1, n_c - 1$, as the primary unknowns.

A fully implicit method (FIM) is used to solve the system of equations (1). The Newton-Raphson method is used, in which a Jacobian (linear)system is assembled and solved for each nonlinear iteration.

Since a composition $\mathbf{z}$ at a given p and T in any grid-cell can be in either a single or two-phase state, the phase equilibrium computation involve a two-step procedure: 1) for any single-phase cell, perform a phase stability test (Michelsen, 1982a) to check if the phase state has changed to two phases or to a different single-phase state; 2) for any two-phase cell, perform flash calculations (Michelsen, 1982b) to compute the phase fractions and compositions at equilibrium.

The flash calculations involve solution of the following nonlinear system

$$z_i - Lx_{i,o} - (1-L)x_{i,g} = 0, \quad i=1,\ldots,n_c, \quad (5)$$

$$x_{o,i}\zeta_{i,o}(p,T,\mathbf{x}_o) - x_{g,i}\zeta_{i,g}(p,T,\mathbf{x}_g) = 0, \quad i=1,\ldots,n_c, \quad (6)$$

$$\sum_{i=1}^{n_c}(x_{i,o} - x_{i,g}) = 0. \quad (7)$$

Here $\zeta_{i,g}$ and $\zeta_{i,o}$ represent the fugacity coefficients of component i in the vapor and liquid phases, respectively. For a given p, T, and z_i, we need to find $x_{i,o}$, $x_{i,g}$ and L. Successive Substitution Iterations (SSI) are usually combined with Newton's method to solve equations (5)-(7) (Michelsen, 1982b).

Adaptive Compositional Space Parameterizations (ACSP)

In Voskov and Tchelepi (2009b), the solution is constructed based on the standard compositional variable set $\{p, \mathbf{z}\}$. That approach is based on the assumption that equation (6) is linear within any triangle so that interpolation of phase fractions $x_{i,j}$ can be used. This linearity assumption limits triangles to a very small size, and that leads to an excessively large number of triangles to cover the compositional space. In Fig. 1, you can see comparisons between a standard EoS-based solution (saturation) for a 1D displacement of a 4-component system and solutions based on tie-line interpolation using two different triangle sizes. The figure shows that the larger triangles lead to larger errors compared with the reference

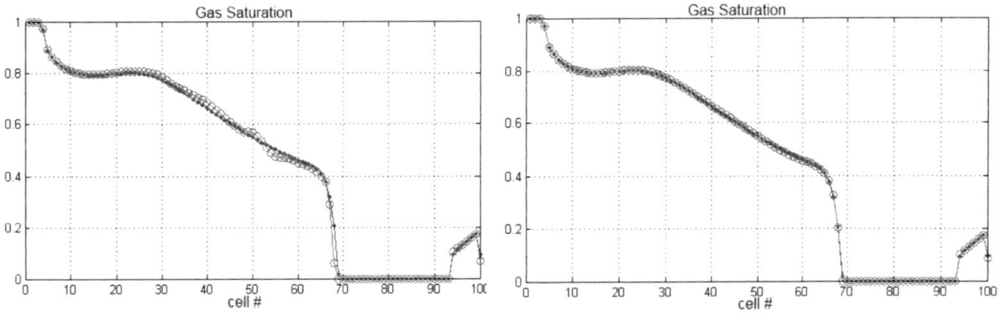

***Figure 1** Representation of saturation distribution for four component system $\{C_1, C_{10}, C_4, CO_2\}$ for large (a) and small (b) size of triangles; here EoS-based solution is shown in solid line, CSP solution in circles.*

EoS-based solution. Moreover, the saturation field obtained with the larger triangles displays small non-monotonic saturation profiles.

Here, we replace the composition vector $\{z_i, i = 1, \ldots, n_c - 1\}$ with a new set, namely, $\{z_j, \gamma_i, i = 1, \ldots, n_c - 2\}$. Here z_j is the composition of the primary component and γ is the vector of the tie-line parameter, that is constant for a particular tie-line. In this work γ is the set of coordinates associated with the center point of the tie-line. Given the tie-simplex based γ parameters, pressure, and temperature, our new phase equilibrium procedure returns the liquid and vapor mole fractions, the phase compositions, and all the necessary derivatives.

For each pressure in the interval of interest, we adaptively construct the phase diagram in parameterized space. For each γ value in the specified set, the tie-lines corresponding to each pressure value in the interval of interest are computed and stored in a table. For any given $\{p, \gamma\}$, we compute the phase compositions for the end points of the pressure interval $[p_j, p_{j+1}]$ using interpolation inside the hyper-triangle. Then, using linear interpolation for pressure $p \in [p_j, p_{j+1}]$, we can find $x_{i,o}$ and $x_{i,g}$. We use Eq. (5) for the primary component z_j when computing the liquid fraction, L, and we define a phase state (liquid, vapor, or two- phase) based on it.

Interpolation in parameterized space

Since a tie-line is a continuous function of p, T and composition z (see Iranshahr et al. (2009)), we can use interpolation in order to compute a tie-line inside a triangle for the pressure interval. To represent a tie-line as a function of the parameter γ, we use the natural-neighbor interpolation scheme (Sibson, 1981), where the n-dimensional ($n = n_c - 2$) tie-line space is triangulated using Delaunay tessellation. For the tessellation of scattered data, linear interpolation of any point inside a hyper-triangle is performed, namely,

$$\gamma_i \quad = \quad \sum_{k=1}^{n+1} \lambda_k \gamma^*_{k,i}, \quad i = 1, \ldots, n \tag{8}$$

where λ_k denote the barycenter coordinates of each vertex of a simplex (n + 1 vertices in n-dimensional space), $\gamma^*_{k,i}$ is the coordinates of the kth vertex. We can write the system of linear equations (8) in matrix notation as:

$$\begin{bmatrix} \mathbf{\Gamma} \\ \mathbf{e} \end{bmatrix} \lambda = \begin{bmatrix} \gamma \\ 1 \end{bmatrix}. \tag{9}$$

Here $\mathbf{\Gamma}$ is a matrix of the coordinates of all the vertices $(n+1)$, and $\mathbf{e}$ is a unit vector, which arises from the constraint on the sum of the barycenter coordinates.

By solving the linear system (9), we can determine the barycenter coordinates of the interpolated point. Since the matrix $\mathbf{\Gamma}$ does not depend on the coordinates of the interpolated point γ, the barycenter coordinate is a linear function of the tie-parameter $\boldsymbol{\gamma}$. Therefore, inside each hyper-triangle, derivatives of $\lambda_k,\ k = 1, \ldots, n+1$ with respect to $\gamma_i,\ i = 1, \ldots, n$ are constants. Knowing the barycenter coordinates of the interpolated point, the mole fractions can be computed as follows:

$$x_{i,p} = \sum_{k=1}^{n+1} \left(x_{i,p}^*\right)_k \lambda_k, \quad i = 1, \ldots, n_c \quad p = \{o, g\}. \tag{10}$$

where $\left(x_{i,p}^*\right)_k$ denote the mole fractions that correspond to $\boldsymbol{\gamma}_k$.

Flash calculation in parameterized space

In order to build supporting points in the tie-line space, we need to compute mole fractions (compositions) for a given p, T, and tie-line parameter $\boldsymbol{\gamma}$. For this purpose, we propose a new flash calculation algorithm in the parameterized Γ-space.

The system of nonlinear equations that needs to be solved in Γ-space can be written as

$$\gamma_i - g_i(\mathbf{x}_o, \mathbf{x}_g) = 0, \quad i = 1, \ldots, n_c - 2 \tag{11}$$

$$x_{o,i}\zeta_{i,o}(p, T, \mathbf{x}_o) - x_{g,i}\zeta_{i,g}(p, T, \mathbf{x}_g) = 0, \quad i = 1, \ldots, n_c \tag{12}$$

$$\sum_{i=1}^{n_c} x_{i,j} - 1 = 0, \quad j = o, g. \tag{13}$$

The definition of g_i depends on how one defines γ_i. In this paper, we use $g_i = (x_{i,o} + x_{i,g})/2,\ i = 1, \ldots, n_c - 2$. For a given p, T and $\boldsymbol{\gamma}$, we need to compute the corresponding $\mathbf{x}_o$ and $\mathbf{x}_g$.

Similar to the standard flash, we can start from phase relations provided by Wilson's equation

$$K_i = \frac{x_{i,g}}{x_{i,o}} = \frac{p_{c,i}}{p} exp\left(5.37(1 + w_i)\left(1 - \frac{T_c}{T}\right)\right), \quad \forall i = 1, \ldots, n_c. \tag{14}$$

Here p_c and T_c are the critical pressure and temperature, w_i is the acentric factor. With the K-values and $\boldsymbol{\gamma}$ parameters known, we can directly calculate $2n_c - 4$ phase fractions:

$$x_{i,o} = \frac{2\gamma_i}{K_i + 1} \quad \forall i = 1, \ldots, n_c - 2, \tag{15}$$

$$x_{i,g} = K_i x_{i_o} \quad \forall i = 1, \ldots, n_c - 2. \tag{16}$$

Next, we need to compute the remaining mole fractions i.e. $x_{i,j},\ i = \{n_{c-1}, n_c\},\ j = \{o, g\}$.

From the definition of the K-values we can write:

$$(K_{n_c-1} - K_{n_c})\, x_{n_c,o} = K_{n_c-1}x_{n_c,o} - K_{n_c}x_{n_c,o} = K_{n_c-1}x_{n_c,o} - x_{n_c,g} \tag{17}$$

which can be rearranged as:

$$\begin{aligned} K_{n_c-1}x_{n_c,o} - x_{n_c,g} &= (K_{n_c-1}x_{n_c,o} + x_{n_c-1,g}) - x_{n_c-1,g} - x_{n_c,g} \\ &= K_{n_c-1}\left(x_{n_c,o} + x_{n_c-1,g}\right) - x_{n_c-1,g} - x_{n_c,g}. \end{aligned} \tag{18}$$

Taking in account equation (13), the last relation can be written as:

$$K_{n_c-1}x_{n_c,o} - x_{n_c,g} = K_{n_c-1}\left(1 - \sum_{i=1}^{n_c-2} x_{i,o}\right) + \sum_{i=1}^{n_c-2} x_{i,g} - 1 \tag{19}$$

Combining (17) and (19), we can easily obtain

$$x_{n_c,o} = \frac{\sum_{i=1}^{n_c-2} x_{i,g} + K_{n_c-1}\left(1 - \sum_{i=1}^{n_c-2} x_{i,o}\right) - 1}{K_{n_c-1} - K_{n_c}}, \tag{20}$$

(21)

The other compositions can be found from (13) and the K-value definition. Next we substitute the $x_{i,j}$ into the fugacities in Eq.(12) and update the K-values using

$$K_i^{(m)} = \frac{\phi_i^{m-1}(x_{i,o})}{\phi_i^{m-1}(x_{i,g})}, \tag{22}$$

where m is the SSI iteration number and ϕ_i is the fugacity coefficient of component i. The SSI iterations are repeated until convergence. SSI can be accelerated by switching to the Newton method at some threshold value of the residual. Such an approach requires linearization of system (11)-(13) for each Newton iteration and updating of nonlinear unknowns $\mathbf{x}_o$ and $\mathbf{x}_g$ based on the solution of the linear system until convergence is reached.

Simulation results

We compare the behaviors of the ACSP-based formulation described here with the conventional EoS-based approach using four- and five-component systems. A brief description of each system is given below.

- System 1: immiscible gas injection for a four-component system. Initial oil is made up of $\{C_1(20\%), CO_2(1\%), C_4(29\%), C_{10}(50\%)\}$ at an initial pressure of 80 bars (above the bubble point). The injection gas is a mixture of two components $\{CO_2(70\%), C_1(30\%)\}$. The injection pressure is 100 bars, which is above the dew point of the injected mixture; the production pressure is 40 bars, which is below the bubble-point pressure of the reservoir fluid. That makes the displacement process highly nonlinear because of the multiple changes in the phase state.
- System 2: immiscible gas injection for a five-component system. The initial oil consists of $\{C_1(1\%), CO_2(1\%), C_4(20\%), C_8(28\%), C_{15}(50\%)\}$ at an initial pressure of 80 bars (above the bubble point). The injection gas is a mixture of five components $\{CO_2(60\%), C_1(30\%), C_4(2\%), C_8(3\%), C_{15}(5\%)\}$. The injection pressure is 100 bars, and the production pressure is 40 bars.

Two different reservoir models are used. Namely,

- Reservoir 1: 1D homogenous reservoir with 500 cells (gridblock size is 10x100x1 meters); permeability is 1D; porosity is 0.2; relative permeabilities are Corey functions with both exponents equal to two. One injection, one production well at the ends of the reservoir. The reservoir temperature is 373 K.
- Reservoir 2: The distribution of permeabilities and porosity is taken from upper layer of the SPE 10 (Christie and Blunt, 2001). The size of the model is [60x220x1] cells (gridblock size is 20x10x70 meters). Relative permeability is Corey with both exponents equal to two. One injection and one production well are placed at opposite corners of the reservoir. The reservoir temperature is 373 K.

Accuracy check

A limited number of hyper-triangles is used to represent the Γ-space. The spacing between the supported points (tie-lines) can be varied in order to achieve the necessary accuracy. Here we compare the results of EoS-based and ACSP-based simulation with hyper-triangles of different size. To define the size of a hyper-triangle, we use the following criteria

$$\|\gamma_i^* - \gamma_j^*\| < \varepsilon, \tag{23}$$

where γ_i^* is the coordinate of the ith supporting tie-line in Γ-space.

Fig. 2(a) shows the solution of Reservoir-1 with System-1 for a large triangle ($\varepsilon = 0.5$). The ACSP-

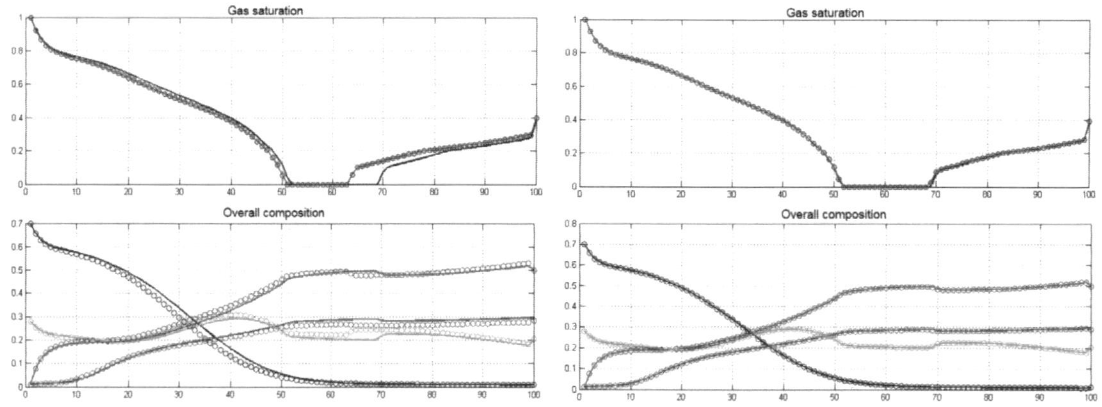

(a) **(b)**

Figure 2 *Representation of saturation and overall composition distributions for four component system* $\{C_1, C_{10}, C_4, CO_2\}$ *with different* ε*. EoS-based simulation is solid lines, ACSP-based simulation - circles.*

based simulation results are shown in circles, and the standard EoS-based results are shown in solid lines. As indicated by the figure, even for very large triangles, the ACSP approach is in good agreement with the EoS-based results excluding a small area close to the displacement front. Simple refinement of the tessellation ($\varepsilon = 0.1$) removes these small errors completely (see Fig. 2(b). The adaptive tessellation results for both of these cases in the Γ-space are shown in Fig. 3(a,b). These figures indicate that we can control the error using tessellation refinement. Note that even when the errors are relatively large, there is no inconsistency in the representation of the physics as opposed to the results in Fig. 2.

2D simulation results of System 1 for the Reservoir 2 model are shown in Fig. 4. The figure shows the saturation distribution, obtained with ACSP-based simulation and difference in saturation after 2000 days. The agreement is excellent except for a few cells around the displacement front. The tesselated Γ-space for this problem is shown in Figure 5. Comparisons between EoS- and ACSP-based simulations for the Reservoir 2 model with System 2 (five component) are shown in Fig. 6. In this case, the dimension of the parameterized space is $n_c - n_p = 3$; hence, the tessellation of Γ-space is represented by tetrahedrons.

Fig. 7 shows the gas saturation (ACSP-based simulation) and difference in saturation after 3000 days. The agreement with EoS-based simulation is excellent.

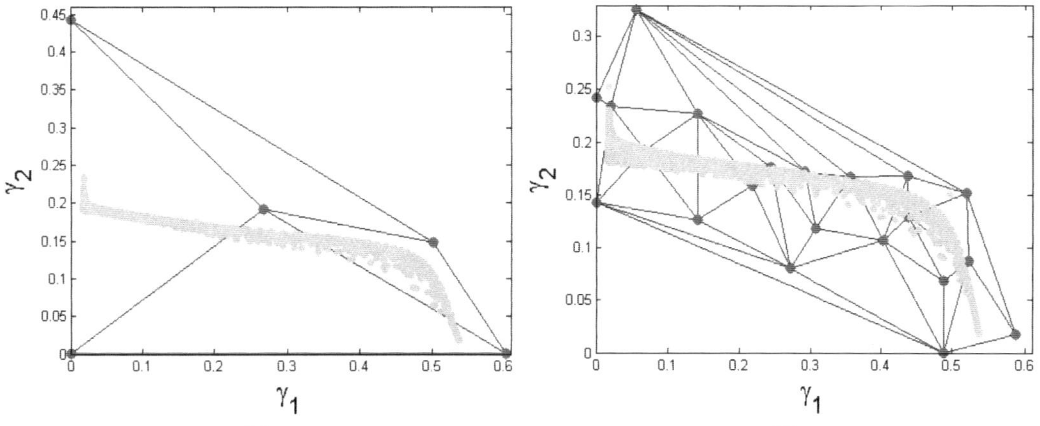

(a) **(b)**

Figure 3 *Triangulated γ space for different tessellations: (a) large triangles and (b) small triangles. Red dots are projections of supported tie-lines. The green dots are projection of the solution.*

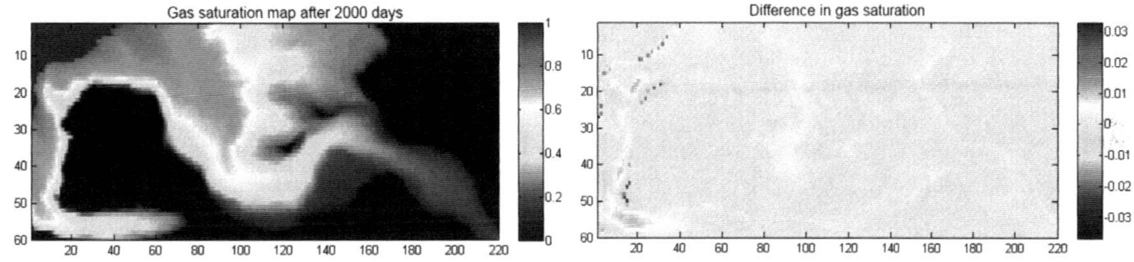

Figure 4 *Saturation and the difference in saturation for EoS-based and ACSP-based simulation after 2000 days for System 1.*

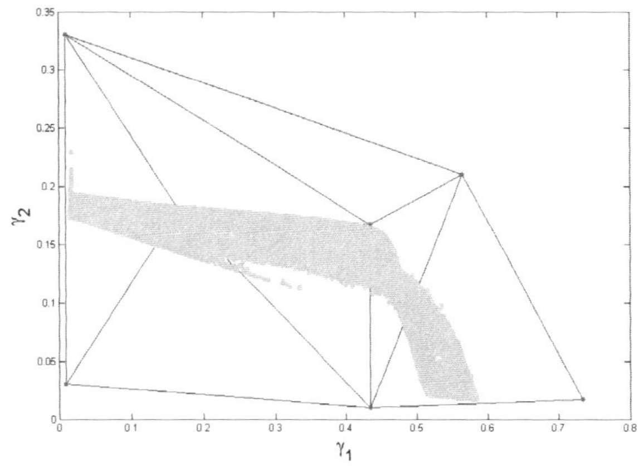

Figure 5 *Triangulated γ space – four-component, two-phase system. Red dots are projections of the supported tie-lines. Green dots are projection of the solution.*

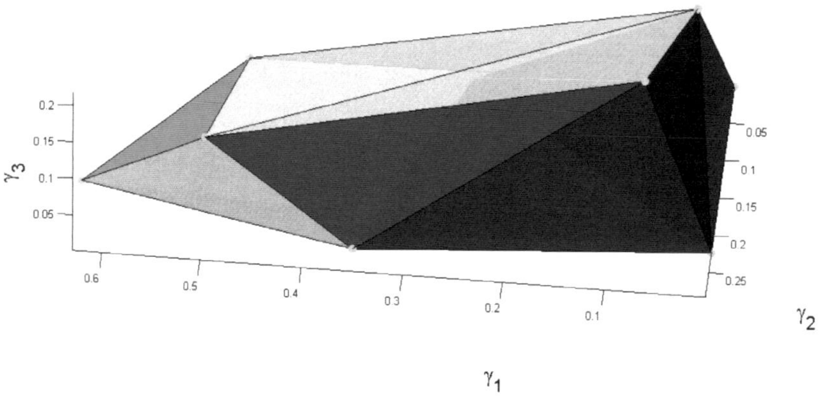

Figure 6 *Triangulated γ space – five-component, two-phase system*

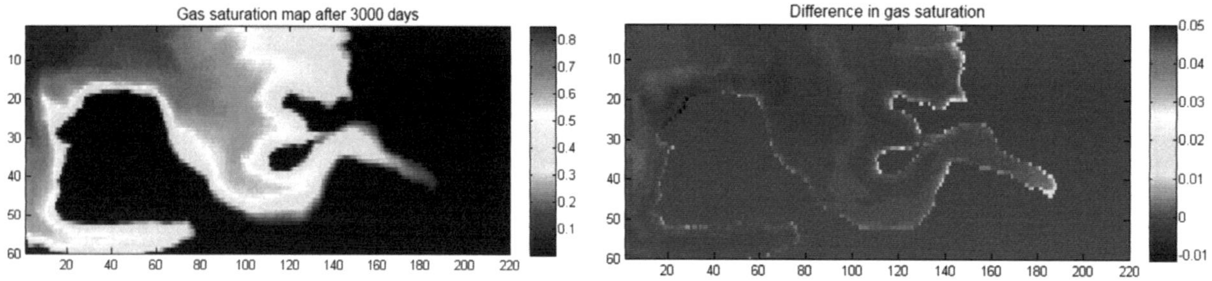

Figure 7 *Saturation and the difference in saturation for EoS-based and ACSP-based simulation after 3000 days for System 2.*

Near-critical representation

The interpolation algorithm described above works well, as long as the solution path is below the hyper-surface of critical tie-lines (Voskov and Tchelepi, 2008). However, in compositional problems of practical interest, part of the solution path may lie in the super-critical region. Here, we discuss how to extend our adaptive parameterization approach to the super-critical region.

The main idea is based on the fact that for a fixed temperature, T, a unique critical tie-line passes through each point in the compositional space (Voskov and Tchelepi, 2009a). Each critical tie-line corresponds to a Minimal Critical Pressure (MCP). As the result, the entire super-critical region is parameterized by critical tie-lines with different MCPs.

Interpolation in the super critical-region can be performed as follows:

- For each supported point, calculate the MCP.
- Find the hyper-triangle that contains the current γ.
- If the current pressure is above the MCP of any vertex, use the critical tie-line defined at these vertexes for the interpolation.
- Otherwise, use the standard tie-line interpolation described above.

This algorithm was tested for several compositional systems of practical interest. In Fig. 8,a you can see an example of interpolation in a large triangle using System 1. We changed the pressure continuously and computed the interpolation errors using a Γ-based flash. Fig. 8,b shows the interpolation error and tie-line length versus pressure. As the pressure increases, the length of the tie-line goes to zero smoothly. The EOS-based tie-line and the interpolated tie-line are in good agreement, even though the size of the triangle is very large. The relative (l_∞) error in the computed phase fraction generally increases with pressure. However, there are no jumps in the pressure values that correspond to the MCP of the supporting point. This fact indicates that the interpolation in the presence of critical tie-lines is reasonably accurate.

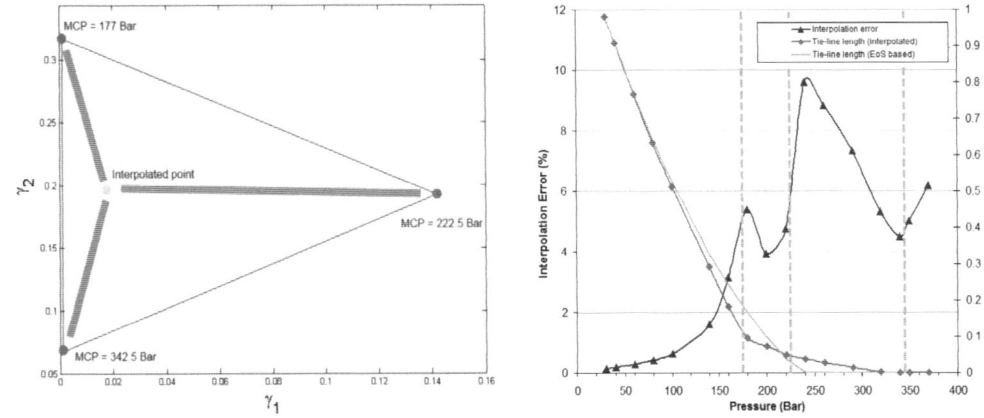

(a) (b)

Figure 8 *Adaptive interpolation for System 1: (a) triangle in the* Γ*-space with corresponding MCP and (b) the interpolation error and tie-line length vs pressure.*

Analysis of the system of conservation law in the hyperbolic limit

It was shown in Entov (1997); Voskov and Entov (2001) that working with the Γ-space can split the full compositional problem into thermodynamic and hydrodynamic sub-problems. The Method of Characteristic theory (Franklin M. Orr, 2007; Entov, 1997) for compositional displacements in one-dimensional systems suggests that solution of the thermodynamic problem is invariant with respect to changes in the properties of the hydrodynamic problem. In practice, it is often the case that even in multi-dimensional problems, projection of the solution in Γ-space follow nearly the same path obtained by the 1D solution (Entov et al., 2001; Voskov and Tchelepi, 2009b).

Based on this fact, we can use characteristics of thermodynamical 1D problems to analyze the behavior of compositional displacements of practical interest. Using Eq. (5) for the primary component j we can find L, and introduce it to the equation for the rest of the components, which leads to

$$z_i = A_i z_j + B_i, \quad i = 1, \dots, n_c - 2, \tag{24}$$

where

$$A_i = \frac{x_{i,o} - x_{i,g}}{x_{j,o} - x_{j,g}}, \tag{25}$$

$$B_i = x_{i,g} - x_{j,g}\frac{x_{i,o} - x_{i,g}}{x_{j,o} - y_{j,g}}. \tag{26}$$

Note that A_i and B_i depend on the γ-parameter only.

It was shown in Entov (1997); Voskov and Entov (2001) that thermodynamical characteristics of compositional problems are controlled by the matrix $\mathbf{R} = \mathbf{B}'(\mathbf{A}')^{-1}$, where

$$\mathbf{A}' = \left[\frac{\partial A_i}{\partial \gamma_j}\right], \quad \mathbf{B}' = \left[\frac{\partial B_i}{\partial \gamma_j}\right]. \tag{27}$$

Our approach allows for separating the pressure and γ variables during simulation, i.e. any function $F(p,\gamma)$ can be represented as $F(p,\gamma) = \Phi(p)\Psi(\gamma)$. In order to analyze the system of conservation laws in the hyperbolic limit, we need to compute the eigenvalues of $\boldsymbol{R}$ for each entry in the pressure table. If one uses the natural neighbor interpolation (Sibson, 1981) for γ points inside a hyper-triangle, the derivatives of the barycenter coordinates with respect to the tie-line parameter γ_i, $\{i = 1, \dots, n\}$ are constants and do not depend on γ. Thus, for each hyper-triangle, derivatives of λ_k, $\{k = 1, \dots, n+1\}$ can be easily computed using Eq. (9) for each pressure value. Next we can substitute these derivatives into Eq. (10) and compute $\mathbf{R}$.

Fig. 9a shows how the eigenvalues for the system shown in Fig. 8 (System 1) change with pressure. In Fig. 9,b the same dependency of eigenvalues on pressure is shown for one of the compositions of System 2. As you can see, all the eigenvalues change continuously with pressure, and this behavior can be a basis for an MOC-based analysis of real systems with large variations in pressure.

Concluding remarks

We presented an adaptive compositional space parameterization (ACSP) method for numerical compositional flow simulation. The methodology builds on the previous work on tie-line (more generally, tie-simplex) based parameterization (Entov et al., 2001; Voskov and Tchelepi, 2009b, 2008). Here, the phase equilibrium relations in the compositional space are recast in terms of tie-line parameters. The approach makes use of the fact that a discrete representation of the tie-simplex space is adequate, and that the solution path associated with compositional displacements are nearly invariant with respect to changes in the hydrodynamic properties.

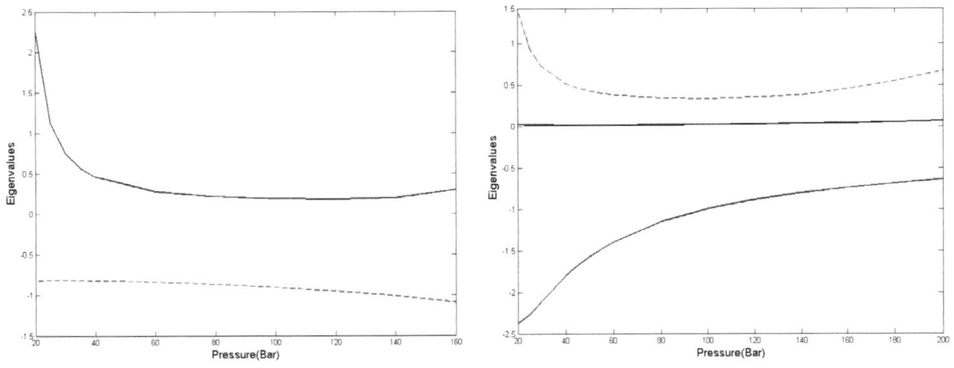

(a) **(b)**

Figure 9 *Eigenvalues vs pressure for: (a) System-1,* γ *corresponds to* $\{CO_2, C_4\}$ *and (b) System-2,* γ *corresponds to* $\{CO_2, C_4, C_8,\}$ *systems.*

We compared the ACSP- and EOS-based methods. The agreement between the two approaches is quite good, even when a small number of tie-simplexes is used to discretize the compositional space. The ACSP approach is consistent with the conservation laws and appears to be quite stable; as a result, the computations are free of oscillations and the accuracy can be controlled to arbitrary levels by refining the tie-simplex discretization. In terms of computational efficiency, for the problems we tested, the ACSP method reduces the cost of the thermodynamic computations quite substantially compared with the standard EOS based approach. Obviously, further testing for problems with highly detailed reservoir models, larger numbers of components, and wide ranges of operating conditions are needed. Moreover, analysis of solution accuracy versus tie-simplex discretization spacing is necessary and is the subject of ongoing research. While further development and testing is needed, the method can be seen as a tie-simplex Γ-based formulation for a general-purpose compositional simulation.

Finally, we note that the the ACSP based formalism makes it possible to split the compositional problem into thermodynamic and transport parts, and this allows for improved analysis of the nonlinear behaviors associated with multi-component, multiphase displacements.

Acknowledgments

Thanks to the SUPRI-B Industrial Affiliates Reservoir Simulation program at Stanford University for financial support.

References

Aziz, K. and Wong, T. [1989] Considerations in the development of multipurpose reservoir simulation models. *proceedings of the 1st and 2nd International Forum on Reservoir Simulation, Alpbach, Austria.*

Bolling, J. [1987] Development and application of a limited-compositional, miscible flood reservoir simulator. *SPE Symposium on Reservoir Simulation, 1-4 February 1987, San Antonio, Texas.*

Christie, M. and Blunt, M. [2001] Tenth spe comparative solution project: A comparison of upscaling techniques. *SPE Reservoir Evaluation and Engineering SPE 72469.*

Entov, V. [1997] Nonlinear waves in physicochemical hydrodynamics of enhanced oil recovery. *Multicomponent flows, Proc. of the International Conference Porous Media: Physics, Models, Simulations, Moscow, Russia.*

Entov, V., Turetskaya, F. and Voskov, D. [2001] On approximation of phase equilibria of multicomponenthydrocarbon mixtures and prediction of oil displacement by gas injection. *8th European conference on the mathematics of oil recovery, Freiburg, Germany*, 4–27.

Franklin M. Orr, J. [2007] *Theory of Gas Injection Processes*. Tie-Line Publications.

Iranshahr, A., Voskov, D. and Tchelepi, H.A. [2009] Theory and practice of tie-simplex parameterization for compositional reservoir simulation. *proceedings AIChE Annual Meeting, Nashville, TN.*

Lake, L. [1989] *Enhanced Oil Recovery*. Prentice Hall, Englewood Cliffs.

Michelsen, M. [1982a] The isothermal flash problem: Part i. stability. **9**, 1–19.

Michelsen, M. [1982b] The isothermal flash problem: Part ii. phase-split calculation. **9**, 21–40.

Pope, G. and Nelson, R. [1972] A chemical flooding compositional simulator. *SPE Journal*, **18**, 339–354.

Sibson, R. [1981] A brief description of natural neighbour interpolation. *Interpreting Multivariate Data*, 21–36.
Todd, M. and Longstaff, W. [1972] The development, testing, and application of a numerical simulator for predicting miscible flood performance. *Journal of Petroleum Technology*, **24**, 874–882.
Voskov, D.V. and Tchelepi, H.A. [2008] Compositional space parameterization for miscible displacement simulation. *Transp Porous Med*, **75**, 111–128.
Voskov, D.V. and Tchelepi, H.A. [2009a] Compositional space parameterization: Multicontact miscible displacements and extension to multiple phases. *SPE Journal*, **14**, 441–449.
Voskov, D.V. and Tchelepi, H.A. [2009b] Compositional space parameterization: Theory and application for immiscible displacements. *SPE Journal*, **14**, 431–440.
Voskov, D.V., Younis, R. and Tchelepi, H.A. [2009] Comparison of nonlinear formulations for isothermal compositional flow simulation. *proceedings of SPE Reservoir Simulation Symposium, SPE 118996*.
Voskov, D. [2002] *Two-Phase Multicomponent Flows through Porous Media*. PhD Thesis, Gubkin Russian State Oil and Gas University.
Voskov, D. and Entov, V. [2001] Problem of oil displacement by gas mixtures. *Fluid Dynamics*, **36**, 269–278.

(302b) Rate Dependent Oil Recovery in Fractured Basement Reservoirs

Justin Wriedt[1], Milind D. Deo[1] and Thomas Doe[2], (1)Chemical Engineering, University of Utah, Salt Lake City, UT, (2)Golder and Associates, Redmond, WA

In fractured basement reservoirs, where oil is primarily stored in fractures, only large aperture fractures are visible through seismic reflection surveys. Considerable reserves may exist in sub-seismic fractures. A discrete fracture network (DFN) simulation study is undertaken to examine the effect of rate on oil and water displacements. The displacement of oil from fractures in waterflooding is controlled by gravity, viscous, and inertial forces. Multiphase displacement in fractures is controlled by fracture dip and continuity of flow paths in the direction of buoyant flow. Higher capacity for flow in seismic fractures can lead to high water production before sub-seismic oil is recovered. High water cut in oil-water basement reservoirs may make oil production economically unrealistic. The effects of threshold capillary pressure, wettability, relative permeabilities, and storage ratios on rate dependent recovery were investigated in this paper. Results indicate that a high rate of production is ideal for optimal oil recovery when time is considered. Comparing pore volumes injected, equivalent recovery is achieved with fewer pore volumes for lower water injection rates. Delayed water production occurs at low rate; however, overall recovery does not increase. Rate dependent relative permeabilities would make low recovery rates more beneficial, however the physical basis for rate-dependent relative permeabilities in oil-water systems is not known at this time. The viscous forces are dominant in high rate simulations. A Darcy flow assumption in the DFN simulations neglects inertial forces which may become important if a Reynolds number > 1 is achieved at higher displacement rates. Results from this study may be used to develop operational guidelines for waterflooding in fractured basement reservoirs. History matching using a DFN approach will improve the understanding of flow interaction in seismic and sub-seismic fractures.

See more of this Session: Simulation of Complex Subsurface Processes II
See more of this Group/Topical: Topical D: Chemical Engineering in Oil and Gas Production and Other Complex Subsurface Processes

Methodology for the selection of small scale LNG technologies for remote gas production

L. Castillo, C.A. Dorao*
Department of Energy and Process Engineering, Norwegian University of Science and Technology, N-7491 Trondheim, Norway
**email: luis.castillo@ntnu.no*

Abstract

Natural gas fields accessible via proven technology are limited demanding that smaller and less accessible gas field should be considered for commercial exploitation. The Liquefied Natural Gas (LNG) could be a solution for this kind of gas reserves, but in the case of remote gas fields with a low productivity, the large scale LNG technologies are neither practical nor economical. The small and middle offshore scale technologies demand different requirements than the ones related to conventional on-land LNG plants. While a thermodynamic efficient process is considered a most important parameter for the selection of onshore LNG plants, other factors need to be considered in the process selection for offshore LNG plants such as safety, compactness, low equipment count, modularization, easy maintenance, reliability, simple operation, low influence of the motion of the vessel, etc. Offshore LNG plants require overcoming many challenges due to the marine environment (winds, currents and waves). The motion of the vessel can affect the process efficiency. In addition, the limited space demands highly compact and light processes. In general, the main characteristics of the Mixed Refrigerant and Expander technologies made them suitable for developing an offshore LNG project, but which is the most convenient choice might depend on the particular scenario.
The most important aspect in this work is that the costs related to the offshore installation are included as a component of the capital cost of LNG project. The plot area required for each technology plays a significant role in the final comparison and this parameter should not be neglected in the economical analysis of ~~future~~ offshore LNG plants.

Introduction

The Liquefied Natural Gas (LNG) is natural gas in liquid state at atmospheric pressure and temperatures around -161 ° C. The volume is reduced a factor 600 times compared to the standard conditions, which allows large volumes of LNG to be transported by sea in refrigerated ships. In the last decades, the world-wide LNG market has growth significantly allowing business opportunities associated with short-term contracts, i.e. spot market, constituting LNG in an important commodity [1].

LNG plants have been designed for high capacities, base-load plant, exceeding 150 million cubic feet per day (Mcfd) of natural gas. The designs of large capacity LNG plants are focused mainly on the exploitation of vast natural gas fields and towards the construction of major facilities taking advantage of economy of scale. However natural gas fields, accessible via proven technology, are limited demanding that smaller and less profitable gas fields should be considered for commercial exploitation. In particular, half gas reserves in the world (6000 trillion cubic feet) are considered stranded, i.e. very little volume of gas, or far away from consumption centers to be economically viable an exploitation project [2].

In the case of remote gas fields with a low productivity, large scale technologies are neither practical nor economical. Different technologies for liquefying of natural gas have been developed for remote gas fields, being the most used the technologies of two and three cycles of cooling, with cascade or propane pre-cooling plus mixed refrigerants schemes. The small to medium scale liquefaction technologies can be divided in two main groups: mixed refrigerants (MR) and expanders based (E) technologies. The first group allows a heat exchanger between the gas and cooler fluid defined as refrigerant or coolant. The second group forces the gas to develop a reverse thermodynamic cycle in one or several steps. These technologies can be differentiated by the refrigerants, heat exchangers or the number of cycles of cooling involved.

The small and middle offshore scale technologies demand different requirements than the ones related to conventional on-land LNG plants. While thermodynamic efficiency of the process is considered the most important parameter for the selection of onshore LNG plants, others factors need to be considered in the process selection for offshore LNG plants. First of all, offshore LNG plants require overcoming many challenges due to the marine environment (winds, currents and waves). The motion of the vessel can affect the process efficiency. In addition, the limited space demands highly compact and light processes. Therefore, selecting an offshore technology should consider several parameters such as safety, compactness, low equipment count, modularization, easy maintenance, reliability, simple operation, low influence of the motion of the vessel, etc. In general, the main characteristics of the Mixed Refrigerant and Expander technologies made them suitable for developing an offshore LNG project, but which is the most convenient choice will depend on the particular scenario [3]. The goal of this work is to discuss a methodology for guiding the selection of small and middle LNG technology based on an economic model where the cost of the plot area is included in the analysis. The proposed methodology can be extended to include other parameters and thus to have the possibility of investigating the effects of several variables in the selection of a particular LNG technology.

Methodology and results

The present work discusses criterion for selecting a small scale LNG technology based on a ranking matrix and an economical evaluation. The first step in this work was to identify a possible scenario for LNG production. The scenario determines the range of the applicability of the technology. The second step was to identify and classify the available technologies. The last step was to perform an economical analysis in order to identify the main parameters that should be considerate for selecting a particular technology. The Figure 1 shows the steps for developing an approach to decision making in LNG projects.

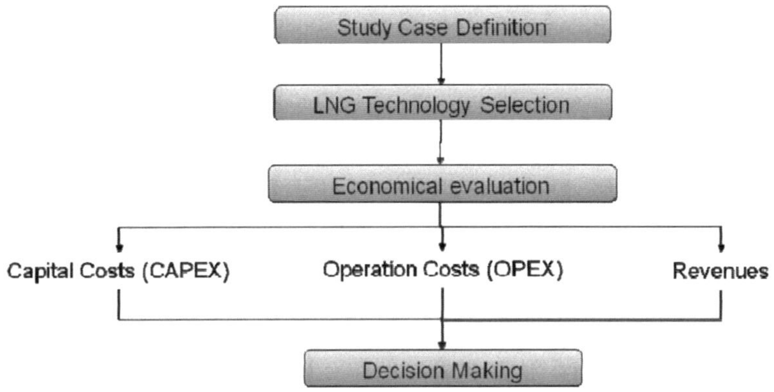

Figure 1: Schematic approach for decision making in LNG projects

Study Case Definition

A generic case is considered where LNG will be located in Venezuela and the main target will be to supply the energy demand of the Caribbean Market, once domestic demand is satisfied. The production capacity might be around 0,5 Mtpa based on typical production rate of the peak shave plants [4].

LNG Technology Selection

Several technologies are available for small scale LNG plants; more details about these technologies could be found in [1,2,5,6]. A first pre-selection of LNG technology could be based on technical analysis as suggested by Castillo et al. [7].

A decision matrix could be constructed in order to evaluate the technologies. The selection of the main parameters could be based on its direct impact in the minimization of investment costs and maximizing efficiency of LNG production. The quality of the information is an important factor that must be took into account in order to obtain reliable results. The weighting stage involves two steps: the construction of the matrix and the weighting to each parameters and sub-parameters. The weights consider the priorities established at the beginning of each particular LNG project. After assigning weights to each parameter and sub-parameters, and defining the appropriate rating, the technologies can be ranked according to the resulting score from the weighted sum of the different parameters measured at the decision matrix. This technique is useful for quick viewing of the strengths and weaknesses of each technology, while allowing comparisons between the options assessed [7].

For simplicity, two groups of technologies are considered: (1) Mixed Refrigerants (MR) with an energy efficiency of 350kWh/tonne, and (2) Expander based with an energy efficiency of 700kWh/tonne. The energy efficiency of the technology can be used for determining the CO_2 production and energy cost.

Economical Analysis

The cost estimates for a LNG project should include many factors, where the capital cost is not the only driver in project economics and must be balanced with the expenditure costs like fuel costs, maintenance costs, and plant reliability. Safety and environmental issues are also important parameters.

The economical analysis for selecting small LNG plant can be related to $\text{CashFlow} = \text{Revenues} + \text{CAPEX} + \text{OPEX}$; where the Cash Flow is the cash into or out of a project during a finite period of time; CAPEX is the Capital Expenditures; OPEX is the Annual Operating Costs and Revenues represents the income due to sale of the products.

Capital cost (CAPEX)

The estimate in the capital cost has been studied during the last decades. Some generic cost estimates are based on historical cost data or simple published estimating formulas [8 , 9]. According to Jamieson et al. [8] investments in the whole LNG value chain, from gas supply through liquefaction, shipping and importation can easily be 4000 - 5000 MUSD for a LNG production of 6 Mtpa. But according to Al-Saadoonat al. [10] , the CAPEX of LNG projects is estimated to be between 200 and 400 USD/tone/year, therefore, LNG plant of 6 Mtpa can cost between 1200 and 2400 MUSD, depending on several factors, including type of technology used, greenfield or expansion project, shipping distance, regasification facility, economies scale, learning curve improvement, and local infrastructure availability.

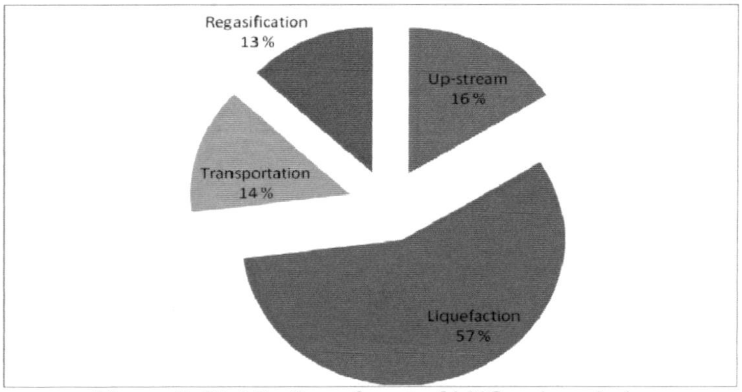

Figure 2. Distribution of cost of LNG value chain

The Figure 2, shows the distribution of cost of LNG value chain. The major cost is related to liquiefaction plant above 50% [9,11]. The liquefaction CAPEX depends on some factors such as technology used, facilities, i.e. gas treatment and storages. In this work, the main focus is on the comparison of the liquefaction technologies and thus only this cost is considered.

According to Barclay et al. [2], the offshore LNG plants has different requirements to the on-load plants, while thermodynamic efficiency is attributed to be the most important criteria for selecting technologies onshore, others parameters can be more important for offshore projects. In particular, the area required impacts significantly in the cost of the offshore installations, and this area is related to the technology.

Figure 3 shows the relation between plot area, power consumption, train capacity and complexity for LNG plant. The Expander technologies require higher power consumption than MR technologies; but they demanded lower plot area and complexity than MR technologies. Moreover, the expander technologies require higher power consumption than MR technologies, but according to Jamieson et al. [8], the liquefaction OPEX can only represent about 5% of the CAPEX. The application of MR technologies for offshore developments requires a reduction of the plot area while maintaining the efficiency of the process, while expander technologies need to improve the thermodynamic efficiency while maintaining a reduced plot area.

Due to the fact the complexity and plot area increase the CAPEX; and CAPEX is an intensive investment, both parameters should be considered into an economical evaluation. The thermodynamic efficiency and plot area play major roles into the economical model because there is a commitment between them on offshore applications.

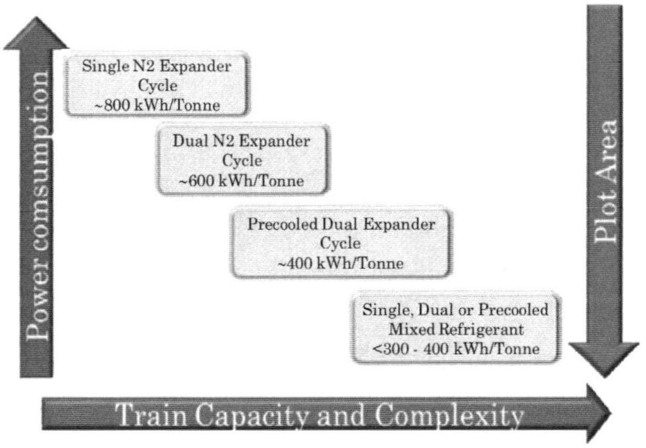

Figure 3. Relation between plot area, power consumption, train capacity and complexity

The CAPEX is assumed to have two main contributions, i.e. $CAPEX = CAPEX_B + CAPEX_A$. $CAPEX_B$ is the CAPEX related to process and can be determined following the work of Garcia et al. [12]. $CAPEX_A$ is the cost related to the cost for installing to process offshore. The estimation of this cost is quite difficult but it can be related to the area required for installing the process at offshore, i.e. $CAPEX_A = Area(P) \times Cost(Area)$, where A(P) is a function of the plot area and depends on the production, and Cost(A) is cost per area unit.

Hence, a simple approach consists in considering that $Area(P) = m \times P + b$; where, m and b depends on the particular technology. Based on this model, a base case was defined, where the plot area as a function of the production is the same for each group of technology; and others three options were defined where the plot area, for MR technology, as a function of the production are higher than expander technology. Finally, the plot area cost is assumed as a fix value and relative to the 25% of $CAPEX_B$, i.e. Cost(A) = 1500 MUSD$/m^2.

Operation Cost (OPEX)

The OPEX of an industrial plant can be resumed in: personnel, energy, maintenance and other costs. In this work, the OPEX can be defined based on make-up of refrigerants; energy costs, CO_2 production, and others fixed cost.

The cost related to CO_2 production is considered as an operational cost. According to world regulations about CO_2 emission, the company should consider the production of CO_2 in their operations. The reference cost under the current European Union carbon dioxide emission quota regimen is about 15 USD/tonne CO_2 [13,14]. Even thought, in the LNG value chain, CO_2 is produced in several stages, in this work only CO_2 related to energy consumption for using small turbine (offshore application) is considered, and it is about 0,7 $KgCO_2$/kWh [15,16].

Hence, the main objective is to compare two groups of the LNG technologies, the others fixed costs, which are the same for both technologies groups, will be assumed as the costs for an LNG plant of 1 Mtpa (about 15 MMUSD/year [12]).

Revenues

In this work, neither NGL or condensates sales are considered. The LNG prices are referenced at Henry Hub Index because the LNG plant is assumed to be located in South America, and thus an average price of 5 USD/MMBtu [3, 17] is considered.

Economical analysis

The economical analysis is based on the following scenario (input data)

Table 1: Input data for the studies cases

Project	
Localization Plant	Offshore
Remote reserve (TCF)	0,1 to 0,5
Production (MTPA)	0,5
Timelife project (year)	25
Discount rate (%)	15
Technology	
MR energy consumption (kWh/ton)	350
Expander energy consumption (kWh/ton)	700
CO2 production (KgCO_2/kWh)	0,7
Market	
$CAPEX_B$ (USD/TPA)	600
Electricity cost (USD/kWh)	0,04
LNG price (USD/MMBTU)	5
CO2 price (USD/tonne of CO_2)	15
Ethane price (USD/tonne)	1400
Propane price (USD/tonne)	620

Internal Return Rate (IRR)

The IRR for each technology corresponding to the base case is presented in Figure 4. The MR technology has a higher IRR than the Expander technology at 15% of the discount rate, when the function of plot area is the same. The difference between the IRR for MR and Expander technologies is about 7%, and thus it is not possible to determine a priori which technology is, economically speaking, more convenient for developing the project.

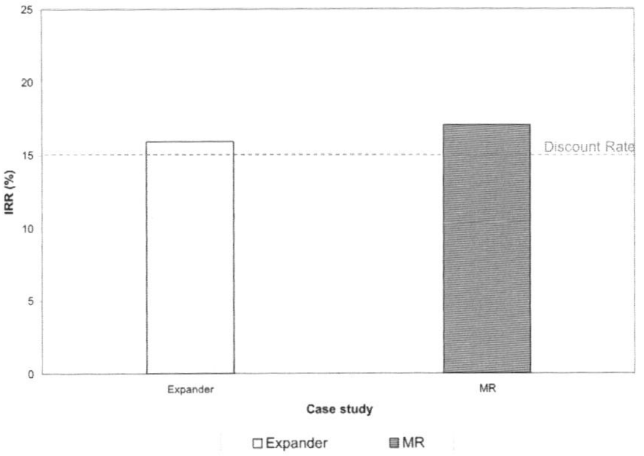

Figure 4. IRR Base Case

Figure 5 displays the economical evaluation for the three options where the plot area for the MR technology changes according to the models proposed. For the three cases evaluated the IRRs are under the discount rate (Figure 5). Considering the Expander technology as a reference, the effect of varying the plot areas for the MR case shows an important impact.

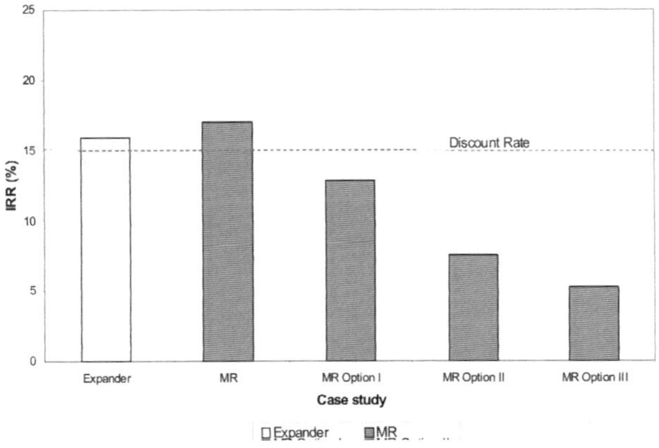

Figure 5. IRRs to the different options

The selection of a particular LNG technology for offshore developments presents the major challenge that thermodynamic efficiency is not necessarily controlling the decision. Parameters like safety, compactness, low equipment count, modularization, easy maintenance, reliability, simple operation, low influence of the motion of the vessel, etc. can impact the overall process and total investment. In the studied case, it was shown that in the economical model the area cost can play a major role in the CAPEX.

Conclusion

In this work a procedure for defining a selection criterion for remote small LNG plants was presented. The approach consists in identifying a possible scenario for remote gas production. The effect of the plot area for offshore LNG plants was considered and in the analysis it was observed that the area plays a major role in economical evaluation of the project. However limited information is available regarding the dependency of the plot area for different technologies and production rates.

References

1 L. Castillo, R. Nadales, C. González and A. Viloria, Liquefied Natural Gas (LNG): Markets and Technologies of liquefaction, Visión Tecnológica journal, 2010 (accepted).

2 M. Barclay, N. Denton, Selecting offshore LNG processes, LNG journal Oct. (2005) 34–36.

3 L. Castillo and C.A. Dorao, Influence of the plot area in an economical analysis for selecting small scale LNG technologies for remote gas production, Journal of Natural Gas Science and Engineering, 2010 (accepted).

4 S. Pérez and R. Díez, Opportunities of monetising natural gas reserves using small to medium scale LNG technologies, in: IGU 24th World Gas Conference, REPSOL, Argentina, 2009.

5 C. Begazo, E. Carvalho and J. Sinoes-Moreira, Small-scale LNG plant technologies, Hydrocarbon World. 2007.

6 T. Shukri, LNG technology selection, Hydrocarbon Engineering, 2004.

7 L. Castillo, R. Nadales, C. Gonzalez, C.A Dorado and A. Viloria. Technology Selection for Liquified Natural Gas (LNG) on Baseload Plant, in: International Gas Conventions & Industrial Gas Exhibitions, Caracas-Venezuela, 2010.

8 Jamieson, D.; Johnson, P. & Redding, P. Targeting and achieving lower cost liquefaction plants, 12th International Conference & Exhibition on Liquefied Natural Gas, 1998.

9 R. DiNapoli. Evolution in LNG Project Costs and Estimation Techniques for New Projects, in: 8th LNG Conference, Los Angeles, 1986.

10 Al-Saadoon, F. & Nsa, A. U. Economic of LNG project, Hydrocarbon Engineering, 2009.

11 A. Fredheim and R. Heiersted. Possibilities for Cost Reduction in Baseload LNG Plant, in: European Applied Research Conference on Natural Gas. 101-114. 1996.

12 E. D. Garcia and F.Sanz, A new business approach to conventional small scale LNG, in: IGU 24th World Gas Conference, REPSOL, Argentina, 2009.

13 J. Jakobsen, G. Tangen, O. Nordbo and M. Molnvik, Methodology for CO2 chain analysis, International journal of greenhouse gas control (2008) 439-447.

14 H.Devold, T. Nestli and J. Hurter, All electric LNG plants, better, safer, more reliable-and profitable, Tech. rep. (2006).

15 H.Herzog, An introduction to CO_2 separation and capture technologies, Tech. rep. (1999).
16 DOE. Carbon dioxide emissions from the generation of electric power in the United States, Tech. rep. (2000).
17 DOE-EIA, Annual energy outlook 2010, Tech. rep. (2010).

(302d) Modeling and Simulation in Tight Gas Sands with Different Fracture Orientation and Studying Impacts of Geo-Mechanical Properties in Tight Gas Sands

Swapnil V. Thakur, John McLennan, Milind D. Deo, Justin Wriedt and Nan Zhao, Chemical Engineering, University of Utah, Salt Lake City, UT

Producing gas from unconventional sources is critical for developing secure sources of energy. In unconventional gas reservoirs like in tight gas sands the permeability values are extremely low. In such cases the natural fractures and fracture networks play an important role in enhancing the production rate by increasing permeability. This paper describes an approach of representing the fracture network and then analyzing the effect of the fracture network on production predictions. The first stage of this process is representing the fracture network as a discrete fracture network. The created discrete fracture network is then introduced into a geo-mechanical simulator. The geo-mechanical simulator is used to create a complex hydraulic fracture in the presence of the existing natural fractures. This system of complex fractures is incorporated into a flow simulator for predicting the production rates and to generate pressure and saturation signatures. This study includes studying the effect of different fracture orientation, fracture spacing on the production values. The simulations are carried to find out how the fractures are created or grown in controlled environments. There is possibility that fracture maybe of irregular shape and may not intersect with the well or with other conductive fractures, but may show conductive features due to the geo-mechanical stress and strains applied on them at reservoir condition. The simulation results for tight gas sands obtained from geo-mechanical simulation coupled with fluid flow are highlighted and discussed in the paper.

See more of this Session: Simulation of Complex Subsurface Processes II
See more of this Group/Topical: Topical D: Chemical Engineering in Oil and Gas Production and Other Complex Subsurface Processes

(302e) An Eulerian-Lagrangian Model to Predict Filter Cake Formation and Expression in Oil Drilling

Jeremiah N. Mpagazehe and C. Fred Higgs III, Mechanical Engineering, Carnegie Mellon University, Pittsburgh, PA

In the oil drilling process, a specially designed slurry forms a filter cake which prevents the premature escape of oil from the well. An understanding of this filter cake's ability to perform in extreme drilling environments is desired. In this work, an Eulerian-Lagrangian model is used to predict filter cake formation and expression. It was developed by combining an Eulerian computational fluid dynamics (CFD) code for the fluid phase with a discrete element method (DEM) model for the particle phase; two-way coupling is achieved between the phases. Individual particle trajectories are determined by fluid drag, electrostatic and Saffman forces. The main results from this work relate to the modeling of the filter cake formation process while varying fluid properties, particle properties, and inlet conditions. Results from the model are validated with experiments.

See more of this Session: Simulation of Complex Subsurface Processes II
See more of this Group/Topical: Topical D: Chemical Engineering in Oil and Gas Production and Other Complex Subsurface Processes

(363a) Relationship Between Kinetic and Flow Parameter Representations in Complex in-Situ Reactive Processes

Jacob Bauman, Chemical Engineering, University of Utah, Cottonwood Heights, UT and Milind D. Deo, Chemical Engineering, University of Utah, Salt Lake City, UT

Modeling reactive-transport processes is complex because it is necessary to calculate heat transfer, multicomponent/multiphase fluid flow, thermodynamics, reaction kinetics, and geomechanics, all models which may have continuously changing parameters. Some examples of these processes are in-situ combustion, pyrolysis, or gasification and CO2 sequestration with geochemical reactions. In this study a novel statistical approach has been implemented to understand the interdependences between reaction and flow parameters.

The approach is applied to in-situ oil shale pyrolysis. The appropriate molecular representation of kerogen, the insoluble solid hydrocarbon in oil shale, is highly uncertain due to its complex nature and variableness across resources. Component lumping schemes have different levels of detail, and the appropriate level of detail is debatable at large scales. Due to the complex nature of kerogen the reaction kinetics of conversion to products are uncertain. The appropriate representation of relative permeability is also a major uncertainty, and is often approximated. Sensitivities of simulated results to these parameters and their interactions are compared using fractional and full factorial experimental designs. Simulations were run with STARS, a thermal reservoir simulator created by CMG.

Ranges of activation energies in hydrocarbon cracking reaction schemes were tested, and it was discovered that specific reaction parameter combinations affect numerical behavior of the simulations. The distance between heating wells and producer wells is a crucial parameter for a successful processing strategy. Heat of reaction for kerogen pyrolysis is insignificant compared to the heat supplied. Relative permeability representation, activation energy distribution for kerogen pyrolysis, and activation energies for kerogen conversion and light oil cracking are dominant parameters.

The statistical approach described exposed parameters that dominate this complex process. Efforts for fundamental understanding and successful production strategies can focus on the most dominant parameters. The statistical methods applied to oil shale pyrolysis can be applied to other complex reactive-transport processes.

See more of this Session: Underground Thermal Treatment of Carbonaceous Materials
See more of this Group/Topical: Topical D: Chemical Engineering in Oil and Gas Production and Other Complex Subsurface Processes

(363b) Flow Simulation with the Lattice Boltzmann Method in 3D Porous Structures of Pyrolyzed Oil Shale Cores Using Multiscale X-Ray CT Imaging

C.L. Lin[1], Jan D. Miller[2] and Ching Hao Hsieh[1], (1)Department of Metallurgical Engineering, University of Utah, Salt Lake City, UT, (2)Metallurgical Engineering, University of Utah, Salt Lake City, UT

The pore structure and connectivity of pore space are important features which determine fluid flow in oil shale during pyrolysis in the production of shale oil. Using a combination of Micro and Nano X-ray computed tomography (CT) and 3D image analysis software, 3D pore networks, kerogen phase, crack structure and flow channels of Mahogany oil shale core samples were characterized before and after pyrolysis at a voxel resolution from several microns to 60 nm. The pore and crack network structure deduced from the images was used in Lattice Boltzmann simulations to calculate the permeability in the pore space. The permeabilities of the pyrolyzed samples of the silicate-rich zone were on the order of milli-Darcies, while the permeabilities of the kerogen-rich zone after pyrolysis were very anisotropic and about four orders of magnitude higher

See more of this Session: Underground Thermal Treatment of Carbonaceous Materials
See more of this Group/Topical: Topical D: Chemical Engineering in Oil and Gas Production and Other Complex Subsurface Processes

(363c) "Cold" Extraction and Liquefaction of Fossil Fuels

L. T. Fan and Shahram Shafie, Green Source Energy, LLC, Austin, TX

Our process for "cold" extraction and liquefaction of various fossil fuels, facetiously named analogous to the "cold" fusion of hydrogen, has been invented based on our serendipitous observations of unexpected phenomena when an essentially green solvent is mixed with powdered coal. It is worth recalling that Salt Lake City is the birthplace of the "cold" fusion of hydrogen, which has totally fizzled. In contrast, however, our "cold" extraction and liquefaction process for fossil fuels has survived third-party verifications by several reputable testing laboratories with flying colors. Moreover, it is well on the way to sustainable commercialization, environmentally and economically, although the domestic and foreign filings of patents for the process have barely been completed.

Various fossil fuels, such as oil sands, heavy crude oil, coal and oil shale, can be extracted or liquefied by our process under or near ambient conditions with a single green solvent. In addition, our process is effected without the aid of additional reagents, such as transport media (water and diluents), catalysts, and surfactants as well as without external supply of large mechanical forces and uncommon forms of energy, e.g., ultrasonic, microwave or nuclear radiation. Apparently, this is hardly the case for any of the available processes for extraction or liquefaction of fossil fuels. These available processes, therefore, tend to be exceedingly expensive to construct and be environmentally unsustainable.

Our "cold" extraction or liquefaction process for fossil fuels is deployable in the subsurface (in situ) as well as on the surface (ex situ). An independent pilot-scale test performed at a leading reservoir engineering laboratory has recovered more than 90% of the residual heavy crude remaining in samples of Berea sandstone plug core with the injection of only 5 pore volumes of our solvent. A third-party verification test carried out at one of the nation's major contract R&D companies has extracted at least 96% of bitumen from Alberta oil sands with an average tar content of 13% in a semi-pilot-scale facility within 5 minutes; the test conditions were 70°C under ambient pressure (1 atm), and the ratio of sample to solvent was 1:1. The residual bitumen on the sands after extraction was determined by an analytical laboratory of Alberta Research Council. Highly encouraging results have also been obtained with a variety of coal by the same company as well as in our laboratory. The extraction of hydrocarbons from coal, however, required about 30 minutes at temperatures approximately between 90 and 100°C with the pressure remaining at 1 atm.

See more of this Session: Underground Thermal Treatment of Carbonaceous Materials
See more of this Group/Topical: Topical D: Chemical Engineering in Oil and Gas Production and Other Complex Subsurface Processes

Thermal Gravimetric – Mass Spectrometry Analysis of Oil Shale

Pankaj Tiwari and Milind D. Deo, Chemical Engineering, University of Utah, Salt Lake City, UT 84112.

Abstract

The fast depleting conventional oil reserves may be substituted by proven vast recourse of petroleum generated source rock known as oil shale. The petroleum source rock contains organic matter laminated with complex mineral matrix. The organic matter is mainly in kerogen form which posses an undefined high molecular weigh structure and upon heating releases the volatile matters. The generated volatile products are both, lighter gases as well as condensable higher boiling point substances known as shale oil, a prospective alternative for conventional oil. The yield of potential oil and composition distributions of pyrolysis process products depend primarily on time-temperature history. The shale oil production yield and improved quality oil from source rock could be enhanced from the knowledge of evolved products composition, competitive reactions mechanism and good knowledge of kinetics triplet. A Thermo-gravimetric analysis instrument with mass spectrometry is potent technique to capture the basic chemistry of the complex process. The evolved components identification based on molecular weight and their thermal behavior at different heating programs provide the in-depth understanding of the products formation and their reaction mechanisms. In this study we report the overall kinetic distribution of Mahogany oil shale pyrolysis using the TGA. The uncertainty quantification is estimated in the distribution of activation energies over conversion scale and simulated model free isoconversional kinetic model (not includes transport effects) is extrapolated to derive the conversion profiles at slow pyrolysis and flash pyrolysis. Single ion monitoring of mass spectra from combined TGA-MS analysis is used to reveals inflections in TGA thermograms during the continuous pyrolysis at different thermal ramps. Identification, relative quantification and thermal behavior of many hydrocarbon products and mineral gases are studied to understand the competitive pyrolysis mechanism of naturally complex petroleum source rock.

Introduction

Oil shale thermal treatment can be performed in both in-situ and ex-situ modes but the produced hydrocarbons distribution is different and yield depends of on other experimental conditions too, temperature, heating rate, pressure and residence time of products in heated zone, etc. The standard experimental techniques such Fischer assay and Rock-Eval can be used to perform the experiments at the normal surface retorting and under high pressure insitu retorting. TGA (thermal gravimetric analysis) is a standard technique to observe the chemical process progress in terms of weight change. The organic matter is mainly in the form of kerogen which posseses an undefined high molecular weight, structure and upon heating releases volatiles. The products generated are mainly in vapor phase and after condensation generate liquid known as shale oil. The potential yield of oil from raw shale depends on the organic maturity of the material that in other way depends on the origin and geological formation of shale. Oil releasing organic matter in shale is broadly classified as naturally present bitumen (very less, max

1wt % of total shale) and kerogen (10-12 wt % of total shale). Some oil shales contain unbound and bound water. Because of tightly bound matrix of the oil shale, significant organic decomposition occurs at a temperature range of 300-550^{0}C. The decomposition of organic matter may even occur at low temperatures, but the rate of the evolution of products will be relatively slower. The understanding of weight loss and continuous evolution of hydrocarbon products can truly be mimicked using faster heating rate. Comparison of the data from relatively two (or more) wide and different heating rates can explain the dependency of product distribution on thermal program.

The organic matter is mainly in the form of kerogen which posseses an undefined molecular weight, structure and upon heating releases volatiles. The products generated are mainly in vapor phase and after condensation generate liquid known as shale oil. The potential yield of oil from raw shale depends on the organic maturity of the material that in other way depends on the origin and geological formation of shale. Oil releasing organic matter in shale is broadly classified as naturally present bitumen (very less, max 1wt % of total shale) and kerogen (10-12 wt % of total shale). Some oil shales contain unbound and bound water. Because of tightly bound matrix of the oil shale, significant organic decomposition occurs at a temperature range of 300-550^{0}C. The decomposition of organic matter may even occur at low temperatures, but the rate of the evolution of products will be relatively slower. The understanding of weight loss and continuous evolution of hydrocarbon products can truly be mimicked using faster heating rate. Comparison of the data from relatively two (or more) wide and different heating rates can explain the dependency of product distribution on thermal program.

The thermal treatment of oil shale produce, condensable vapor, non-condensable gases and spent shale. The yield and composition distributions of different products pyrolysis process depend on time-temperature history. The potential gaseous products start evolving when petroleum source rock reaches some definitive temperature and continue as the temperature rises. The generated volatile products are both, lighter gases as well as condensable higher boiling point substances known as shale oil. The shale oil is valuable products and considered a prospective alternative for conventional oil. The yield of total shale oil production from source rock may be enhanced from the knowledge of evolved products compositions and competitive reaction mechanism. It has long been recognized that the fraction of organic carbon converted to oil depends on the H/C ratio of the oil shale. Thus, the ultimate analysis (elemental compositions) of the raw oil shale and spent shale is necessary to figure out and support the theoretical understanding with broad sense to the mechanism of the products formation. Because oil shales have differing origins and are found in differing geological environments, and the compositions are different (H/C ratio, organic % and mineral matter's matrix) it is not surprising that they behave differently when subjected to varied pyrolysis conditions.

The major hurdle in the development of commercial production of oil shale is a variability of parental oil shale compositions, both organics and minerals. The rock formation varies, at depth and horizon in terms of composition. The comprehensive understanding of oil shale pyrolysis requires the detail information of the raw material composition, transport of heat, rate of products formation, distribution of products, interaction and alternation of minerals and organics, resistance of mineral matrix during products release, etc. The presence of some minerals may catalyze the hydrocarbon generation and the resulted pore formation may reduce the possibility of secondary reactions, such as coking, cracking, etc. Thus, multiscale thermal treatment of oil shale, experiments with different size of cores may reveal the insight of thermal heavier of rock substance, resistance distribution offered by pore formation and variability in the products rates and compositions during pyrolysis.

The compositional analysis of the products obtained can be estimated in many ways. GC (Gas chromatography) analysis is usually performed on the samples collected after condensation and provides the information on overall products independent of the thermal behavior. Combined GC-MS (Gas-chromatography- mass spectrometry) analysis is a similar and powerful technique, which provides the identification of the components in the sample. DSC (Differential scanning calorimetry) gives the heat value information of the samples. FTIR (Fourier Transformed Infrared spectrometry) provides the methodology about compositions identification and wax appearance temperature. HPLC (High performance liquid chromatography) is also a similar technique, which quantifies the products in groups (SARA- saturates, aromatics, resins and asphaltene). XRD (X-ray diffraction), SEM (Scanning electron microscopy) and EDS (Electron dispersive spectroscopy) are used for the characterization of solid products (qualitative and quantitative analysis of the minerals in the original and reacted assemblages) from the experiments.

A Thermo-gravimetric analysis instrument with mass spectrometry is potent technique to capture the basic chemistry of the complex process. The evolved components identification based on the molecular weight and their thermal behavior at different heating programs provide an in-depth understanding of the possible products and their reaction mechanisms. The single ion monitoring of water vapor, methane and carbon-dioxide in mass spectra over temperature scale reveals inflections in TGA curve during the continuous pyrolysis and could be used appropriately for distinction of organic and mineral decomposition sections. Identification, relative quantification and thermal history of many hydrocarbon products from pyrolysis of mahogany oil shale were obtained from TGA-MS instrument. The identified organic compounds were classified based on the common structure and bonds, like, aromatic, alkane, alkene, alkyne and some PAH (poly aromatic hydrocarbons). The rational comparison of different products is used to understand the pyrolysis mechanism of naturally complex petroleum source rock.

Experimental setup

Elemental Analysis

The elemental analysis is appropriate to classify the oil shale based on the ratio of hydrogen to carbon (H/C) and oxygen to carbon (O/C) before performing the experiments. The raw oil shale samples were obtained from Utah Geological Survey department and were crushed and screened to 100 mesh (1.49 E – 4 meters) size particles. The CHNSO (carbon, hydrogen, nitrogen, sulfur ad oxygen) elemental analyses of the sample were performed using LECO CHNS-932 and VTF-900 units and results are summarized in the Table 1 with standard deviation (10 runs for each). The elemental analysis shows that studied oil shale is Type I on Van Krevelen classification diagram in the term of H/C (1.1) and O/C (0.67) atomic ratios.

Table 1. Elemental analysis of the oil shale sample

Oil Shale Sample	*Average %*	*Stdev*
Carbon	17.45	0.26
Hydrogen	1.6	0.08
Nitrogen	0.53	0.06
Sulfur	0.18	0.04
Oxygen	15.69	0.79
H/C % (molar)	1.1	
O/C % (molar)	0.67	

Thermo Gravimetric Analysis-Mass spectrometry

The crushed 100 mesh-size samples of Mahogany oil shale were dried for four hours to remove any moisture; if present. There was no significant weight loss during drying, thus the samples are used as received after screening to 100 mesh size. To study the thermal behavior of oil shale pyrolysis (N_2 environment) products, a TGA instrument (TA Instruments Q-500) coupled with mass spectrometry was used for the entire temperature range of kerogen decomposition. The particles size used along with other conditions, was specifically selected to allow the products to leave the heated chamber of TGA with no transport effects. Nitrogen was used as sweeping media to perform the experiment in inert environment. Total flow rate of nitrogen was 100 ml/min (90ml/min as purge and 10 ml/min as balance gas). The possibility of secondary reactions during the decomposition and because of significant residence time of products in the reactor (heated chamber of TGA) was taken care using the high sweep gas flow rate. It is assumed that the total sweep gas flow rate is enough to carry over the entire vapor out from the chamber and prevent any vapor phase decomposition and condensation of the products. The combined TGA (thermal) and MS (monitoring of products) were performed at four different heating rates (Table-2). The TGA furnace chamber outlet was connected to MS instrument through heated (150C) capillary column line to prevent any condensation of the products, and it is assumed that products are in original form as they produced. The vacuum is maintained in the MS to reduce the residence time of the sample. The total number of components to be identified by single ion monitoring of mass spectrometry was changed in each experiment while the basic components such as methane, water, carbon-dioxide and light hydrocarbons (aklanes, olifins and aromatics) were fixed in all experiments to analysis the comparative and competitive reaction mechanism of naturally complex oil shale materials. The experiment at 5C/min thermal rate was repeated adding more compounds in ion monitoring catalog.

Results and Discussion

The TGA is a standard technique, which records the weight change data over the time-temperature history. The weight loss of oil shale is mainly because of the thermal treatment of solid material and due to this phenomenon the evolved products are mainly in the vapor phase. The sufficient flow rate of the purge gas carry those vapor out of the heated chamber and it can be assumed that the resident time is not enough to allow the products to participate for secondary reactions. The combined TGA-MS experiments are performed at different temperature programs. Table-2 summarized the experimental condition and information obtained from TGA weight loss curve.

Table 2: Experimental conditions

Heating rate	Final temperature	Total compounds	Weight, mg	%wt loss @250C	%wt loss @500C	%wt loss @600C	T max, C
0.5C/min	650C	43	22.54	0.34	9.38	20.14	401.03
1C/min	600C	56	13.10	0.65	9.6	15.99	413.09
5C/min-1	960C	22	13.34	0.59	10.76	13.2	442.61
5C/min-2	600C	56	13.26	0.46	10.3	11.29	443.41
10C/min	650C	41	19.05	0.63	11.28	12.47	456.93

TGA produces overall weight loss thermograms during the pyrolysis and facilitate with the products in vapor phase as they generates to be indentified in attached mass spectrometry. The evolved product gases were identified in single ion monitoring response of mass spectrometry based on molecular weight. The list of the molecular weight was created based on the GCMS analysis of the shale oil produced in the lab and results published in the literature. That includes some basic key compounds that may be used as sectional division criteria for the TGA curve as well as represent the basic chemistry when molecules change their form and explain the physical phenomena of thermally activated process products, are H_2O, CO_2 and CH_4. For example, evolution monitoring of the methane may be an indication when organic decomposition occurs and presence of CO_2 and water peaks at different section on thermal scale may be indication of organic and mineralogical transformation. The water sign at the early stage may explain the presence of bounded and unbounded water in the sample. The other compounds molecular weight were used are normal hydrocarbons, could be straight chain, branched, olefins, cyclic, aromatic and PAH.

The combined TGA-MS analysis of any process products is a powerful approach to study the reaction mechanism and further kinetics development but on the other side, limited in various ways specially for complex material like oil shale where many unknown compounds may be in the products. There is a lag on the temperature scale in TGA weight loss data and corresponding ion current signal in mass detector. Most of the organic weight decomposition in the TGA data are completed around 500C (first peak in organic decomposition temperature range). This may be because of the time required for product gases to travel from TGA furnace to ion detector through capillary tube under vacuum. Second possible reasons could be that some components were still being formed at higher temperature while the total weight loss was not significant and corresponding not was contributing in the DTG peak. The TGA weight loss and the area of the components in that specific section cannot be compared because of two primary reasons. First, there could be more compounds, which may be present in the sample but were not taken in to consideration and secondly, the ion current response of components is not on a different scale and cannot be quantified and compared to other runs. But the mass spectrometry ion current for a single run allows us to do the relative quantification of the competitive reaction mechanism, such as the thermal behavior and relative amount by computing the area of one component and compare to the others. The maximum rate temperature from TGA is for combined weight loss and cannot be compared to individual ion monitoring maximum signal temperature for a specific component obtained from mass spectrometry. That's why the temperature range was extended till 600C for components relative quantification in this study. Some potential components molecular weight may conflict with other structural compounds. For example nonane and naphthalene (128 amu) and, carbon-dioxide and propane (44 amu) share the same molecular weight and make it impossible to distinguish based on single ion response. The presence of oxygen, nitrogen and sulfur may release the products with oxygenated, nitrogenated and sulfur attached hydrocarbons. In this study the GCMS information is used to assign the name of components. Mostly components are straight statured and unsaturated hydrocarbon and aromatic structures were found in GCMS total ion spectra (TIC) of shale oil sample. The possible products distribution of oil shale pyrolysis and their corresponding molecular weights along with the onset points for some key components are summarized in Table-2. The first experiment was performed at 5C/min (denoted by 5C/min-1) from ambient to 960C and total 22 compounds molecular weights were added for ion monitoring. The results suggested that all of the organic compounds traces are before the 500C and carbon dioxide and H_2S because of mineral decompositions started after 600C. Thus, to focus the analysis in the organic decomposition section, further experiments were design upto 600C with different heating rates. The experiment at 5C/min is repeated (5C/min-2) to insure the instruments reliability for components identification and therograms (Figure-2). The single ion monitoring compounds were added in the list based on the results in the earlier runs. The total 56 compounds were added in 0.5C/min and 5C/min-2 runs while the total limit of the MS instrument is 63 compounds. The MS instrument is also constrained with the limitation on highest molecular weight, 300 amu. Thus, the products of higher molecular weight (more than 300 amu) can not be detected with TGA-MS instrument.

Analysis of TGA Thermograms

The total weight loss of oil shale from ambient to 500C at different heating rates was found to be in the range of 10-12% of total initial weight. The higher heating rates shift the peak to higher temperature and total organic decomposition happens at relatively lower temperature than lower heating rates. This is because of time-temperature history; at low heating rate, the sample spends more time at small temperature range. The derivative of the TGA thermograms is shown in figure-1. The offset has been set differently to see the clear view. All the run don't have the peak in the samples at early stage except 1C/min has tiny peak in the temperature range of 80 to 180C. This could be either because of water loss or molecular rearrangements. The experiments at 5C/min for wide temperature range (960C) have the second large peak after 600C, because of the mineral decomposition.

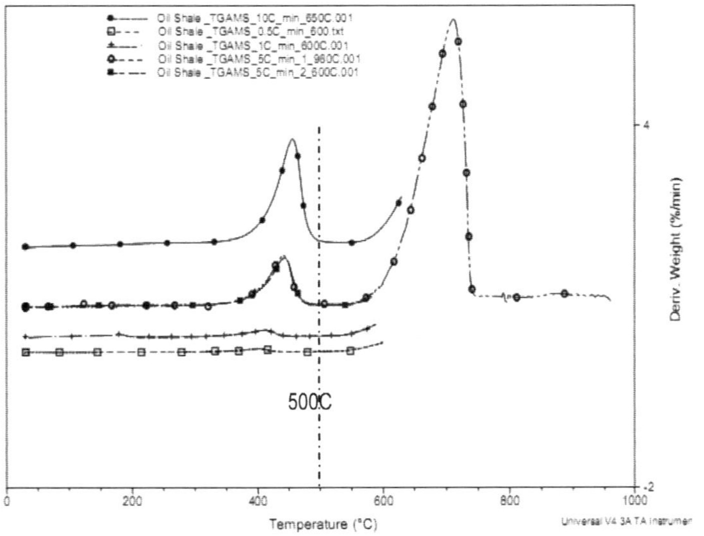

Figure-1: TGA thermograms at different thermal programs

The organic decomposition temperature scale may be divided in the three sections. The weight curve has been divided in three sections, A1-30 to250C, section A2 -200 to 500C and section A3- 500 to 600C. This information can be obtained from table by subtracting the different section weight loss from ambient to that temperature. As can be seen from the tables and also a clear indication in the figure that there is no significant weight loss in the section A1 that indicates a very small or no water presence in the samples, TGAMS curves also shows the water absence below 200C. The weight loss is very less in this section and could be cause of the molecular arrange of the sample as the temperature increases at low heating rate. Section A2 shows the significant weight loss and this weight loss is mainly because of the organic matter decomposition in the samples. The third section (A3) has the significant weight loss information but this could be because of the mineral matter decomposition, at slow rate, better time-temperature relationship this weight loss is more and mostly attributes to mineral decompositions or rearrangement in mineralogy of the sample.

Analysis of ion monitoring – mass spectrometry

Single ion monitoring allows the inflections in the DTG thermograms to be unambiguous identified based on molecular masses. The compounds evolution depends on the heating rates as well as the programs of thermal history. Traces results of the common compounds at different heating were similar except the intensity of signal and thermal behavior. At low heating rates products come out earlier as well as don't have the significant presence in terms of peaks as they spend more time at the low temperatures and continuously evolved. The detector intensity of ion current depends on some factors, first is amount of the sample, second the rate the particular ion in reaching to the detector. In the case of the low heating rates the peaks of identified compounds are present, but not with high rates as for the higher heating rates. The similar trends as we observed on the TGA weight derivative curves were seen in the mass spectra. At low heating rates the peaks are in the same region but the intensity of most of the compounds signals are too low to get the sharp peak onset points. Similar to TGA thermograms, the MS single ion monitoring response for the products generated from raw material at different heating rates like the derivative peaks intensity is higher at higher heating rate (10C/min). Figure-2 shows the single ion response of mass spectrometry for different molecular weight compounds. The data were plotted in the lumped format to get the better view. Alkanes, olefins (alkenes and alkynes), benzene based aromatics and poly aromatic hydrocarbons along with inorganic gases are plotted in lumped format. The Figure-2 reveals that in all hydrocarbons the lighter products get released earlier. At higher heating rates the alkene compounds shows the peaks at low temperature; this could be because at higher temperature these unbounded compounds get released. Higher molecular weight compounds at respective lumped group need more temperature to get formed.

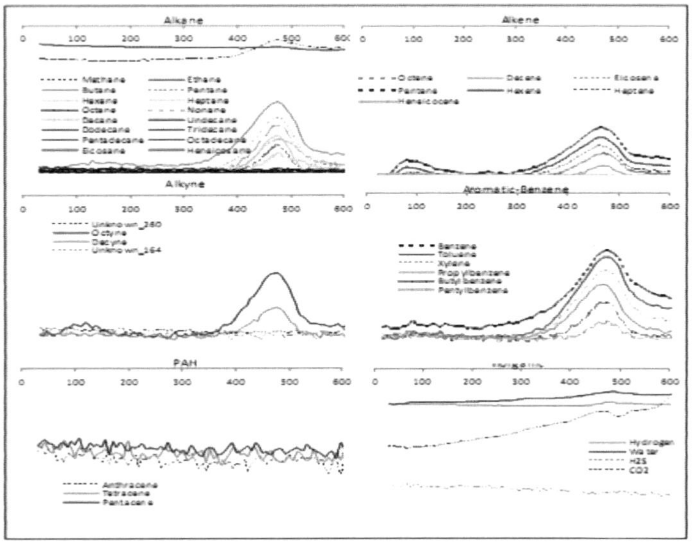

Figure-2: Classified relative components single ion monitoring over thermal scale at 10C/min ramp rate

The PAH compounds have lot of noise and hard to identified are they present in the sample or not and where are the onset points for them. The baseline shift is also different for all the compounds, especially light organic and inorganic gases. Single ion monitoring of the basic key components, water, methane and carbon-dioxide release hypothesis can be explained based on the signal response at different heating rates, shown in figure-3. The reproducibility at 5C/min is good agreement but compare to heating rate (10C/min) the signal are showing the low intensity.

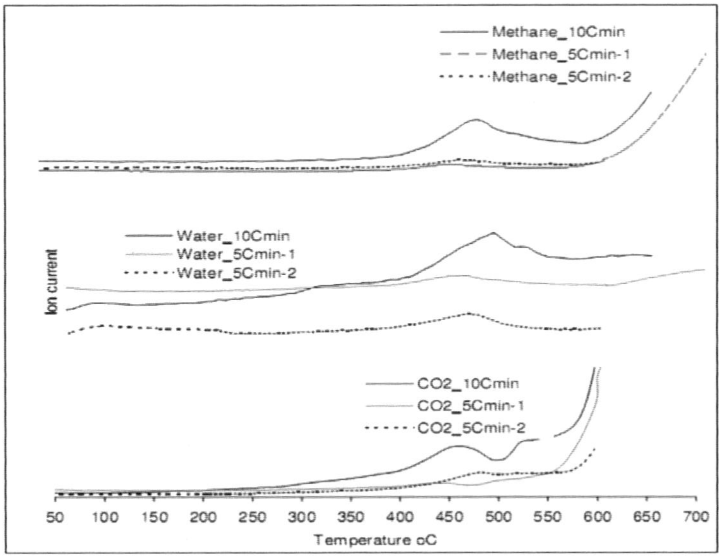

Figure-3: Single ion monitoring signal of methane, water and carbon dioxide at different heating rates

There is a steady stream for all these three compounds before the organic decomposition section (less than 300^0C) that confirms the absence of naturally present water in the sample. All these three compounds have significant signal strength in the organic evolution temperature range, because during thermal loss of organic oxygen as water and through some mineral dehydroxylation, methane as a light hydrocarbon and carbon dioxide as the products of oxygen and carbon contained material pyrolysis. The CO_2 and methane peaks are having the more strength at higher temperature, this may be because the mineral decomposition release CO_2 and decomposition of remain higher molecular weight hydrocarbon goes under cracking and release the light gas, methane.

The comparison of the single ion monitoring for different compounds could be a better illustration of thermal behavior and the same compound could be compared at different heating rates for the thermal response. Figure-4 shows the some alkanes, alkenes and aromatic compounds thermal behavior at two heating rates. The y axis is an ion current response of the mass spectrometry for diffent compounds based on the molecular weight. This cannot be used for quantitative analysis and omitted from all the graphs. Aromatic compounds signal appears at low temperature then respective carbon number hydrocarbons (Benzene vs hexane and decane vs butybezene). The maximum rate temperature for specific compound at different heating rate is almost identical and the traces follow the similar path. Start onset point for olefins of a specific hydrocarbon is at early temperature then the respective normal alkane. The difference on the TGA weight loss curve over temperature and the spectra of some compounds obtained from mass spectrometry show the deviation on the temperature, signals are spread above 500^0C.

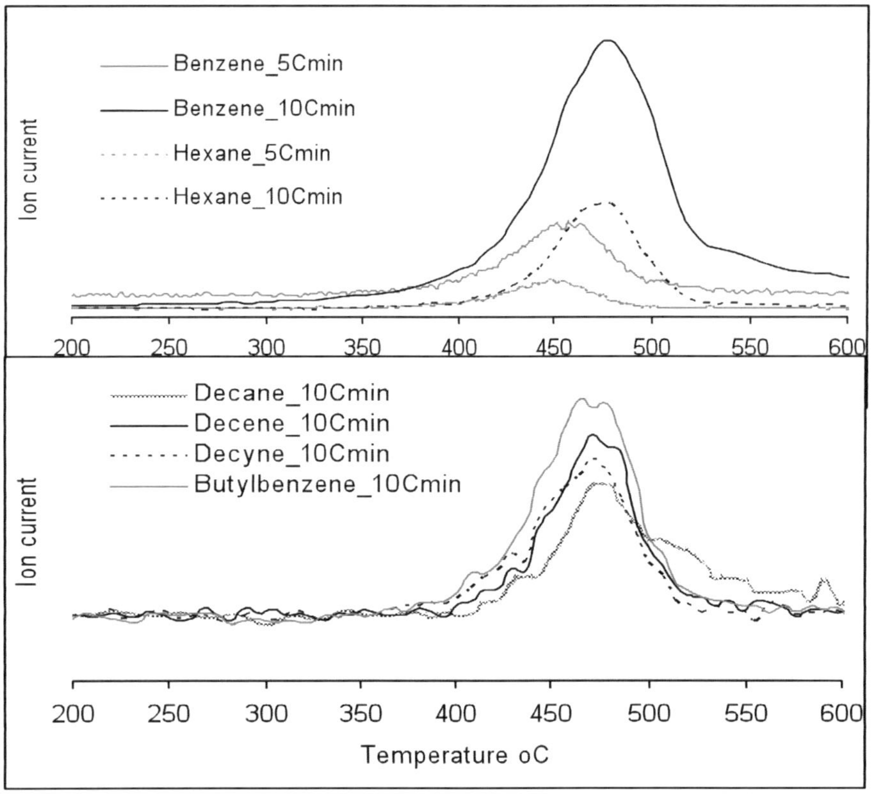

Figure-4: Comparison of the key components on thermal scale at different heating rates

Higher boiling points compounds, can not make an educated judgment just based on their evolution temperature range about their origins but the relative thermal rate dependency with other runs and relative quantitative presence could be obtained from the information obtained at different temperature rate TGAMS runs. The area under the single ion monitoring over temperature range is calculated by trapezoidal method. The correction for shift in the baseline was deducted using the minimum value of the particular component (rectangle of minimum value over temperature range). The area integral is used for the relative analysis of the components evolved. The area under the curves could be a good basis to compare the ratio information for the compounds at particular heating rate and at different heating rates. The area is divided in two section, before the organic decomposition temperature (30^0-250^0C) and organic pyrolysis temperature (250^0-600^0C). Some compounds complete the peak before 600^0C, but it is assumed that the return to base line and does not contribute extra in 600^0C section. Some compounds second peaks, like methane, water and CO_2 started before because of the mineral decomposition, thus were treated separately as not be compared with rest of the hydrocarbon compounds. Another reason is there baseline shift is different then the rest. Figure-5 shows the distribution of some components at 5C/min in two different section. For the hydrocarbons, the section of 30-250C could be ignored or used as another correction of the noise in the signal as they not get released in this section but the area is estimated because of the noise in the signals. At %C/min aromatic (benzene) dominate the other compounds and highest hydrocarbons are mainly present in the products (less than13 carbon number). The small window shoes the absence of key components (or very small percent) as compared to in the second section.

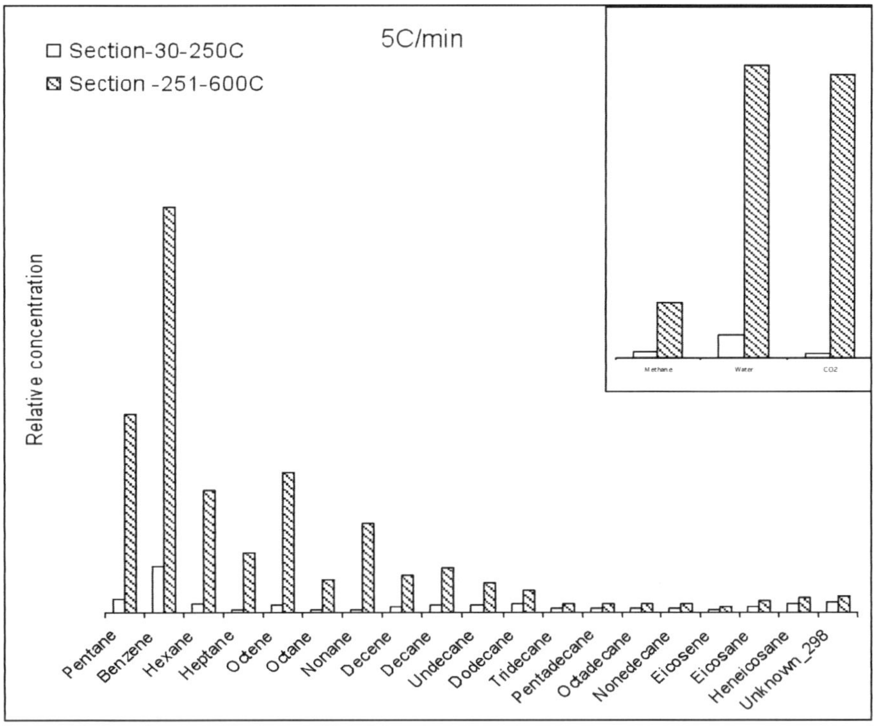

Figure-5: Relative area of ion current response of the components at 5C/min ramp rate

Conclusion

The TGA weight loss information along with the single ion monitoring of mass spectrometry provides the components identification and their thermal behavior at programmed heating rate. The obtained signal from MS identifies specific gas components during thermal analysis. The onset points (start, end and maximum rate temperatures) for a component identified my single ion monitoring reveal not only the presence but also their thermal activities at different time-temperature history. The monitoring of water and CO_2 at significant level and on the temperature scale distinguishes the combined processes such as organic decomposition and mineralogical changes. Different hydrocarbon signals at ramped thermal programs provide the understanding about the products composition, temperature dependency and the relative quantification. The source rock material is very complex and it is not possible to determine the formation route of the products. But, based on the results it could be concluded that the source rock includes a variety of hydrocarbons, like the presence of substantial aromatics and molecules are attached through different light hydrocarbons (saturated molecules and olefins). During thermal treatment, or more specifically when a source rock reaches a definitive temperature at specific heating rate, the bond connection breaks and low molecular weight products formed. At fast heating rates temperature onset points shift to higher scale side and produce more aromatics and olefins. The limitation of TGA-MS data don't allow to perform absolute quantified analysis of the products and relate it to source material. But, this kind of analysis may play a vital role in correlating GC and GCMS analysis of products obtained from pyrolysis of source rock in standard apparatus, such as Fischer-Assay and Rock-Eval. The combined analysis may permit the absolute quantification with temperature distribution of the products and further may be used in the development of a comprehensive kinetic model.

(363e) Reservoir Based Hydrothermal Reforming of Hydrocarbons

Yousef Alshammari, Imperial College London, London, United Kingdom and **Klaus Hellgardt**, Chemical Engineering, Imperial College London, London, United Kingdom

Enhanced Oil Recovery (EOR) operations may offer a potentially clean and economical method for hydrothermal production of syngas or hydrogen as useful product or energy vector. EOR operations usually employ in situ combustion (using oxygen or air), steam injection and/or CO_2 injection. In situ combustion has been studied for a number of years, e.g. the THAI process invented in 1993 by M. Greaves. In this process an air injection well is used to combust heavy oil and thus provides local heat to reduce oil viscosity. The oil is then driven towards a producer well (toe to heal). It is suggested that approximately 80% ooip recovery may be accomplished with the additional benefit of some in situ oil upgrading. Air injection results in the release of an equal amount of CO, CO_2 and hydrocarbons. This of course begs the question whether the return gas, which would be used to provide the required electricity for the compressors, could be converted into syngas/hydrogen and thus lead to the production of a clean energy vector. Indeed, is it conceivable to employ, for example, redundant production wells as hydrothermal reactor systems with the explicit injection of oil, steam and air/oxygen and thus remove the need for above ground reformers? In the present work a basic parametric equilibrium study has been carried out on the reactions between crude oil and EOR fluids (air, steam, and CO_2) using HYSYS and employing the embedded Gibbs Reactor model. This model allows the thermodynamic analysis of the effect of the molar flowrates of injection fluids (ratio of injection fluid to oil), reservoir pressure, reservoir temperature and water cut on the composition of the produced syngas. Only a few similar investigations have been reported in the literature (Ye et al., 2009, Ersoz et al., 2003, Jeon, 2008). These studies do not normally include crude oils containing sulphur compounds, nitrogen compounds and asphaltenes. Adiabatic steam/air reforming has been compared with adiabatic CO_2/air reforming and adiabatic partial oxidation in terms of product distribution, temperature, and compression load. The goal was to identify conditions under which hydrogen production is maximised (Figure 1). It was found that steam, containing 20% O_2, may potentially provide the highest yield of H_2, the lowest yield of CO_2, with reasonable compression load compared with air or CO_2. The main challenge of air injection is the huge compression load, due to its high N_2 content, which can reduce its cost-effectiveness. On the other hand, the main challenge with CO_2 reforming is the huge rise in the CO_2 compared with the very low H_2 yield, which may add additional cost for separation of CO_2 above surface. In situ sequestration of CO_2 underground before reaching the wellhead might be feasible.

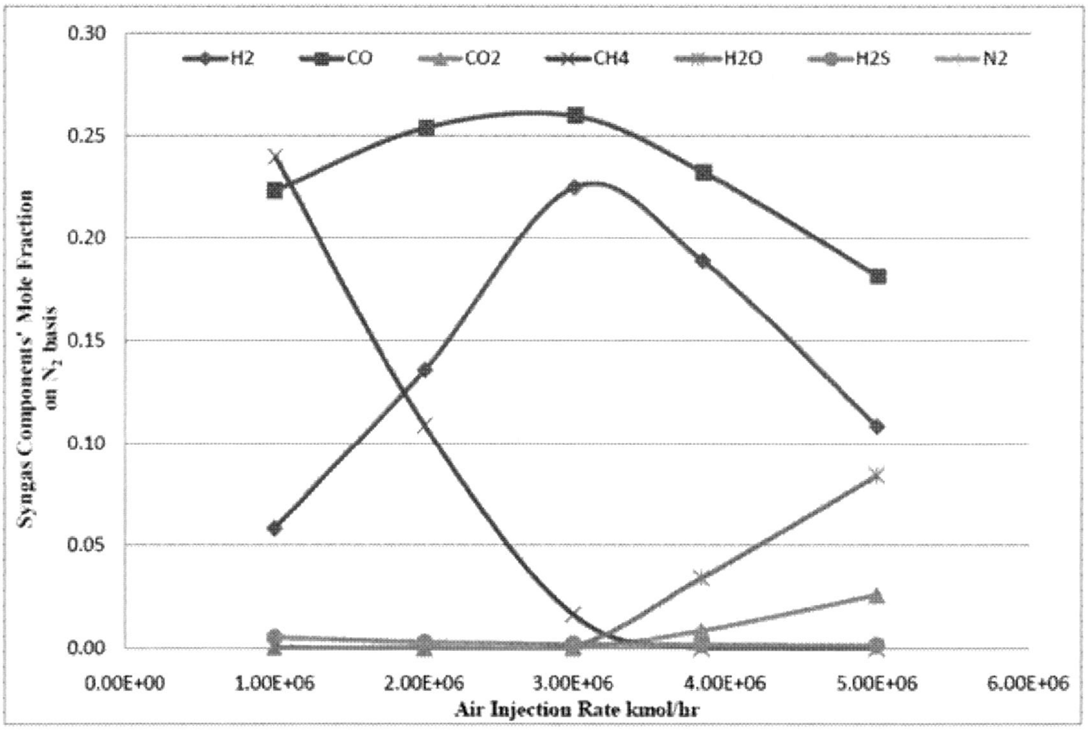

See more of this Session: Underground Thermal Treatment of Carbonaceous Materials
See more of this Group/Topical: Topical D: Chemical Engineering in Oil and Gas Production and Other Complex Subsurface Processes

(363f) Development of CFD-Based Simulation Tools for in-Situ Thermal Processing of Oil Shale/ Sands

Benjamin J. Isaac and Philip J. Smith, Chemical Engineering, University of Utah, Salt Lake City, UT

In-situ technologies are currently being explored because of their potential for reducing the environmental footprint of oil shale/sands development. However, the first generation technologies have proven to be energy-intensive, and many unknowns remain relative to optimal heating strategies, potential contamination of groundwater, and achievable production rates.

Reservoir simulation tools are typically applied to in-situ production processes. However, we are simulating a modified in-situ process where a distribution of rock size and orientation (important due to directional permeability of the rock) exists in the production bed. The rocks are heated by convective currents through the channels of the rubblized bed. Our approach is to apply CFD-based simulation tools to this modified in-situ process. The system consists of a clay-lined volume filled with rubblized oil shale that is heated by pipes fired with natural gas burners. We have employed a suite of commercial software tools, including EDEM, Matlab, Gambit, and Star-CCM+.

We have determined that a simulation with the actual geometry of the pieces of shale will provide more accurate results than treating the rubblized shale volume as a continuum with a defined permeability. DEM Solutions product, EDEM, a discrete element modeling software that is used for modeling particle movement, was used to simulate the filling of an empty bed with pieces of oil shale. This simulation models the interactions between particles and geometry as they fall into the bed. The simulation resulted in a bed full of thousands of pieces of oil shale that can be used as a representation of the actual rubblized bed. The EDEM simulation provides data about the position, size, and rotation of each particle.

Matlab was then used to write a journal file that defines the three-dimensional layout of the particles. The journal file was run in the CAD software Gambit to create the shale geometry. Gambit produces a parasolid transmit file which is a standard geometry file type that can be used in Star-CCM+. Implementation of the correct geometry representation in Star-CCM+ requires being able to represent boundary interfaces between pieces of shale and fluid.

We demonstrate the efficacy of this set of tools for the rubblized shale system, using the tools of validation and uncertainty quantification to perform a consistency analysis between simulation results and experimental data.

See more of this Session: Underground Thermal Treatment of Carbonaceous Materials

(389a) Modeling Diffusion Coefficients of Carbon Dioxide in Water for Carbon Sequestration

J. Wambui Mutoru, Alana Leahy-Dios and Abbas Firoozabadi, Chemical Engineering, Yale University, New Haven, CT

Abstract

The increase in atmospheric concentrations of carbon dioxide (CO_2) is believed to be a key player in global warming, necessitating identification of viable technological options for its capture and storage. Perhaps, the most promising of these options is injection into saline aquifers whose merits include large sink capacities, and enhanced storage integrity upon CO_2 dissolution. This growing interest in CO_2 injection calls for a better understanding of CO_2 diffusion – quantified by infinite dilution (denoted as D^{∞}) and Fickian (denoted as D) diffusion coefficients – into water and brine. However, there is neither sufficient experimental data for $D^{\infty}$and$D$, nor unified models for accurate predictions. We propose a new model for calculating $D^{\infty}$for CO_2-water mixtures. The model takes into account the temperature dependence of the dipole moment of water and the polarizability of CO_2,and fits experimental CO_2-water data at low and high pressures with an accuracy of 4.9%. Remarkably, the proposed model also accurately predicts $D^{\infty}$for binary mixtures of methane (CH_4) and hydrogen sulfide (H_2S) in water where solute polarizability is comparable in magnitude to that of CO_2. Both CH_4 and H_2S are often present as impurities in CO_2-injection streams. Moreover, we present – to the best of our knowledge – the first predictions of composition-based D for CO_2-water over the temperature range 298.15-413.15 K, and pressures up to 50 MPa.

See more of this Session: Carbon Dioxide Injection in Underground Formations
See more of this Group/Topical: Topical D: Chemical Engineering in Oil and Gas Production and Other Complex Subsurface Processes

Effect of rock composition on mineralization in sequestration

Prashanth Mandalaparty, Graduate Research Assistant, University of Utah.

Chemical Engineering Dept, University of Utah, 50 S. Central Campus Dr., MEB, Rm. 3290, Salt Lake City, UT 84112

Email: Prashanth.mandalaparty@gmail.com Phone: 3365099709

Dr.Milind Deo, Professor, Chemical Engineering, University of Utah.

Chemical Engineering Dept, University of Utah, 50 S. Central Campus Dr., MEB, Rm. 3290, Salt Lake City, UT 84112

Email: milind.deo@utah.edu Phone: 8016520451

Dr.Joseph Moore, Research Assistant Professor, Energy and Geo-science Institute, University of Utah.

Energy and Geo-Science Institute, University of Utah, 423 Wakara Way suite 300, Salt Lake City, UT 84108.

Email: jmoore@egi.utah.edu Phone: 8015856931

Abstract

The reactive behavior of pure CO_2 with limestone, sandstone, arkose and peridotite was examined in this study. The experimental apparatus consists of series of high-pressure reactors with pure CO_2 as feed gas at 100^0C to evaluate the dependence of kinetics and mineralogical changes on rock composition. Arkose was observed to exhibit the highest tendency of participating in these reactions, which can be attributed to the geochemical complexity of its initial mineral assemblage. Layers of calcite were seen growing on the surface of the arkose. Analcime deposits are almost omnipresent either occurring as large connected aggregates or as deposits on surfaces of other minerals. Ankerite and calcite deposition were observed as amorphous mass intergrown with starting minerals. Continuous dissolution of limestone was seen with the release of CO_2 gas, indicated by the increasing pressure in the reactor (formation of a gas chamber), which occurs due to the lack of any source of alkali to buffer the solution. With sandstone, there was slight increase in pH due to dissolution of feldspars. The rate of carbonation of feldspars is pretty slow compared to the dissolution. Hence the precipitation of carbonates at the measured time scale is not evident. In peridotite experiments, carbonation of peridotite forming calcium and magnesium carbonates along with serpentine was evident. Hence, arkose has the geochemical complexity for permanent sequestration of CO_2 as carbonates and is an ideal choice. The study provides useful laboratory data when considering CO_2 sequestration in different geologic The brine chemistry results for the experiments with different initial rock compositions are presented in this report. These results complement the changes in the rock chemistry.

Introduction

Carbon dioxide is the most abundant GHG (64%). The atmospheric concentration of CO_2 has increased from 280 ppm during the preindustrial period to 385 ppm with about half this increase having occurred since the mid 1960's. In the United states majority of the CO_2 emissions are from power plants which account for about 40% of the total emissions (IPCC report 2007) and it was identified before by Holloway et al 1997. Carbon dioxide sequestration appears to be an important potential method by which emissions into the atmosphere can be reduced. In this method anthropogenic carbon dioxide is disposed into geologic formations including saline aquifers, depleted oil reservoirs (CO_2 EOR) and unmineable coal seems (Enhanced coal bed methane recovery). These formations are widely available and also in close proximity to majority of the point emission sources (Holloway 1997). Injection of CO_2 deep underground is particularly promising because deep sedimentary formations have the potential to retain CO_2 in the subsurface for thousands to millions of years (Bachu 2002). In the United states the capacity of deep saline aquifers is greater than any other geologic formation with very high estimates (Bergman and Winter 1995) and they are also found within close proximity to power plants (Bachu 1994). Aquifers with salinities exceeding 10,000mg/l total dissolved solids are excluded by the U.S. Environmental Protection Agency as underground sources of drinking water (Xu et al2005). Hence these form a primary target for the eventual disposal of CO_2.

The temperature in these aquifers (100C-200C) varies greatly depending on the depth and also the local geothermal gradients (Bachu 2002). Also CO_2 is injected at depths greater than 700m to ensure that it stays in the super critical form because its critical point is approximately 304 K and 1073 psi (Span and Wegner 1996) and also is soluble in water. The notion of CO_2 disposal in aquifers has been discussed in the literature with specific aquifers as target like The Netherlands (Lohuis 1993) and the Alberta Basin, Canada (Gunter 1993 Bachu 1994 Perkins and Gunter 1995, Gunter 1996 and 1997 Bachu 1996).

Geologic formations, which are artificially charged with CO_2, are called Carbon repositories (Kaszuba et al 2005). Once injected CO_2 being more buoyant than water will rise until it hits an impermeable membrane or low permeability seal. This is called structural or stratigraphic trapping of CO_2. Then depending on a rate controlled by several factors such as the rate of CO_2 injection, the rate of CO_2 dissolution into the pore water, the surface area available for the reaction and the rate of diffusion of the CO_2 into the pore water away from the pore water- CO_2 interface, CO_2 dissolves in the formation water (Allen et al 2005). This process is termed solubility trapping. Then CO_2 is trapped as a discontinuous phase (either in supercritical form or as dissolved CO_2) in the pores of formation leading to capillary trapping. Then CO_2 forms carbonic acid, which dissociates to form carbonate and bicarbonate ion leading to ionic trapping of CO_2. Finally these anions react with the cations resulting from the dissolution of primary minerals, due to decreased pH, which in turn leads to precipitation of carbonates permanently sequestering CO_2 (Bateman et al 2005). Hence logically, mineral precipitation can be termed sequestration where as other mechanisms can be termed as storage (Kaszuba et al 2005).

Less importance has been given to mechanisms by which CO_2 is actually trapped as immobile mineral phases. Kaszuba et al (2003) carried out a study to determine the extent of fluid rock interactions in addition to carbonate mineral precipitation that may occur in an experimental system that simulates geologic storage and sequestration of CO_2. They reported the precipitation of magnesite, siderite and analcime. Kaszuba et al (2005) then carried out a study to analyze the effect of CO_2 injection on fluid rock reactions and also assess the effect of these geochemical interactions on the integrity of the cap-rock. Shale was used to model the aquitard. Such analyses for the integrity of cap-rock had been done previously by Lindeberg (1997) using simple numerical models (Darcy equation and Fick's law). They attributed the leakage of CO_2 to gravity migration with subsequent release through sub vertical fractures and faults. A few experimental studies examined geochemical reactions in a saline aquifer in response to CO_2 injection. Gunter et al 2000 and Perkins and Gunter 1995 carried out numerical geochemical modeling studies that incorporated kinetic laws and some studies combining experiment and modeling, in which dissolution of silicate minerals in brine and precipitation of carbonate are reported. The dominant state of carbonic acid with respect to pH plays an important role in governing the factors that lead to formation of carbonates in aquifer (soong et al, Drever 1988, Drucken miller et al 2005). Drucken miller et al reported the formation of calcite at pressures of around 600psi and at 150C. Rosenbauer et al reported the affect of presence of SO4- to the geochemical reactions and also the changes it brings about in the solubilities and the porosities of the repository.

O'Connor et al 2002 investigated the feasibility of aqueous mineral carbonation for mineral sequestration of CO2. They used a brine consisting sodium bicarbonate and sodium chloride with a mineral reactant (magnesium silicates such as olivine or serpentine) and dissolved CO2 into this slurry by diffusion through the surface and dispersion through within the aqueous phase, This resulted in the precipitation of magnesite from the carbonation of the magnesium silicate minerals. The optimum conditions for this process were determined and also a method for mineral preparation for the process was also developed. In their further study it was found that these slow geological processes could be accelerated by increasing the reactive surface area, increasing the activity of carbon dioxide in the solution, inducing imperfections into the crystal lattice high energy attrition grinding and also thermally activating the mineral.

One of the main factors governing this complex geological environment is the initial mineral assemblage. It has been stated in the literature that the type of rock in the reservoir also governs the type of the trapping mechanism of CO_2. For example in a limestone reservoir the principal method of trapping is ionic trapping and structural trapping. Inorder to verify this hypothesis the analysis of these sequestration reactions with both the rock and brine chemistry has been provided in this study.

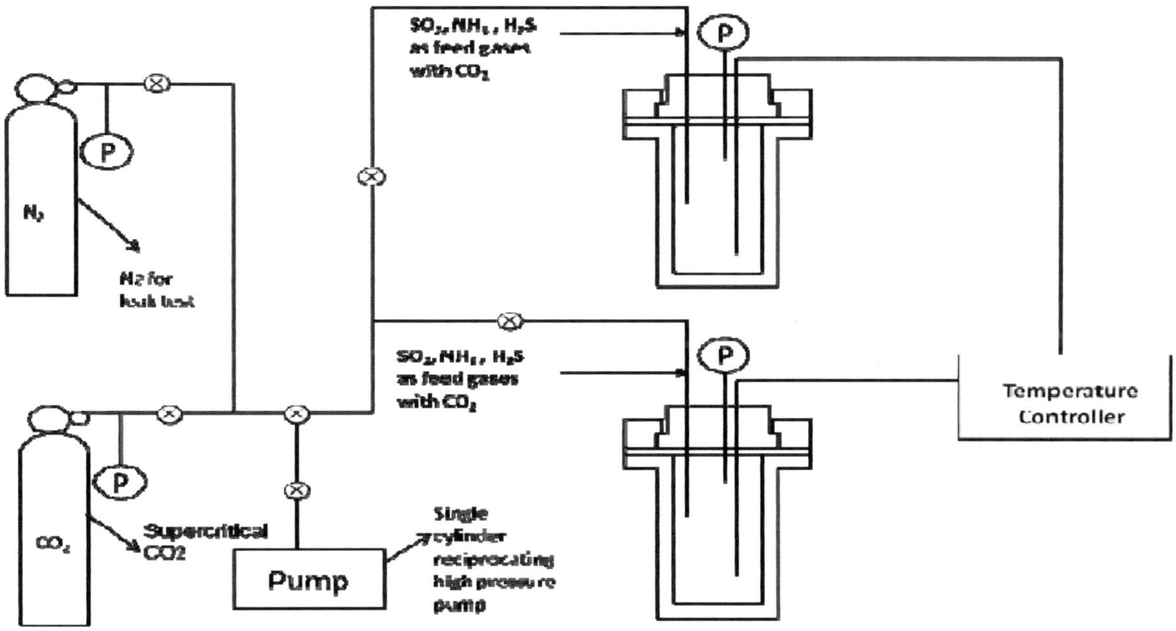

Figure 1: Schematic diagram of the experimental setup.

Experimental setup

The experimental apparatus (Figure 1) consists of series of stainless steel reactors made of 316-grade stainless steel, which were rated for 4000psi at 600C. The reactors were pressure sealed with high-pressure SWAGELOK fittings. The reactor has provision for retrieving the rock sample without disturbing it through a detachable cap (Swagelok fitting) at the bottom. The CO2 was pressurized into the reactor using a single cylinder high pressure reciprocating DB Robinson pump. The flow of CO2 into the reactor was controlled using high-pressure needle valves. Syphon CO2 (CO2 in supercritical form) was used for these reactions. High purity nitrogen was used to pressure test the entire setup at 3000psi. The temperature was controlled using a bench top temperature controller with SPECVIEW as the interface via K-type thermocouples. Brine was prepared from laboratory grade NaCl with its initial composition shown in Table2. The ions in addition to Na and Cl might be due to the impurities in the salt used. High purity nano-filtered de-ionized (DI) water was used to prepare the brine samples. The rock analysis was carried out using X-ray diffraction XRD, Scanning electron microscope and EDS analyses. The advantage of using such apparatus over the previous experimental described in Seyfried et al which was adopted by Kaszuba et al (2003,2005) and Rosenbauer et al (2005), is that we can correlate the changes in mineral chemistry with the changes in brine chemistry at each stage the sample is collected, which gives us a comprehensive picture of the geochemical interactions taking place. The changes that the occur due to sampling ie., the shift in the equilibrium between CO2 and brine, the geochemical reactions occurring at the brine-mineral interface are also not disturbed. Since this is a closed system there is a one-one relationship between brine chemistry and mineral chemistry. The changes we see in brine chemistry should correspond to the changes we see in mineral chemistry. This experimental setup helps us in evaluating that.

Experiment conditions

The different sets of experiments are tabulated in Table 1. All the rocks were crushed to 100μm and mixed to construct the arkose. All the minerals were angular to circular in shape and were ranging from 40-100μm in size.

Expt	**Rock used**	**Temperature**	**Pressure**
Set A	Limestone	100^{0}C	2000 psi
Set B	Sandstone	100^{0}C	2000 psi
Set C	Peridotite	100^{0}C	2000 psi
Set D	Arkose	100^{0}C	2000 psi

Table 1: Experimental conditions

All the experiments were carried out with 3g of rock. For the Arkose, the initial sample was prepared by mixing equal proportions of rock (0.5 grams each). After the experiment, the sample was retrieved from the reactor by detaching the bottom (wherein the sample rests in a cap) and carefully taking the sample out without disturbing the precipitants on the surface. The sample was dried overnight at 50^0C, temperature not high enough to alter the crystal structure of any minerals. The sample is then retrieved, cooled down and divided into equal parts for XRD and SEM/EDS analyses. The quantitative estimates of the minerals before and after the experiments were obtained by XRD. Since very small amounts of precipitates were formed, any attempt to quantify produced solids proved difficult. Hence the changes in the initial mineral assemblage and the mineral assembly after the reaction were calculated. These changes can be corroborated with the precipitated minerals seen in the SEM/EDS analyses and a reaction mechanism leading to the precipitation of new minerals can therefore be postulated. After the experiments the brine samples were diluted and filtered using Whatman 40 filter paper (retention capacity of 8μm) in a vacuum filtration setup. The sample was then divided into sub equal fractions for all cation and anion analyses. The sample was acidified a little by addition of sulfuric acid to prevent any further precipitation. All the dissolved cations were analyzed using ICPMS and dissolved anions by ion chromatography. Table 2 gives the initial composition of the brine

Na	**Mg**	**K**	**Ca**	**Al**	**Mn**	**Fe**	**Ba**	**Si**	**S**	**Cl**
mg/L	mg/L	mg/L	mg/L	μg/L	μg/L	μg/L	μg/L	mg/L	mg/L	mg/L
23032	1	<6	<4	<8	<1	54	<2	0.4	<6	26542

Table 2: Initial composition of the brine

Results and discussion

Set A: Experiments with pure Limestone

Initial XRD and SEM analysis (Figure 2) of limestone revealed it to be 99.4 calcite and 0.6% quartz . Hence the sample was relatively pure limestone. During the experiment, which was carried out for a period of 42 days the pressure increased to 2060 psi after 17 days and 2090 psi after 42 days. This can be attributed to a four-step process as indicated in the order of the reactions below.

(Eq 1)$CO_2 (g) \leftrightarrow CO_2 (aq)$

(Eq 2)$H_2O + CO_2 \leftrightarrow H_2CO_3$

(Eq 3)$H_2CO_3 \leftrightarrow H^+ + HCO_3^-$

(Eq 4)$CaCO_3 + H^+ \leftrightarrow Ca^{2+} + HCO_3^-$

The carbon dioxide dissolves in water to form carbonic acid (Eq1 and Eq 2). This leads to a decrease in pH. pH decreased from a value of 5.4 to about 4.9 after 28 days and 4.7 after 42 days. Because of this increased acidity calcite, which is sensitive to changes in pH undergoes dissolution to release CO_2 (Eq 3 and Eq 4), which causes the increase in pressure inside the reactor. The absence of any silicate minerals to buffer the brine solution in the reactor rules out the possibility of any precipitants and also mineral sequestration of CO_2. The dissolution patterns are almost omnipresent (Figure 3) through out the sample when the SEM analysis was carried out.

Figure 2: SEM analysis of initial calcite

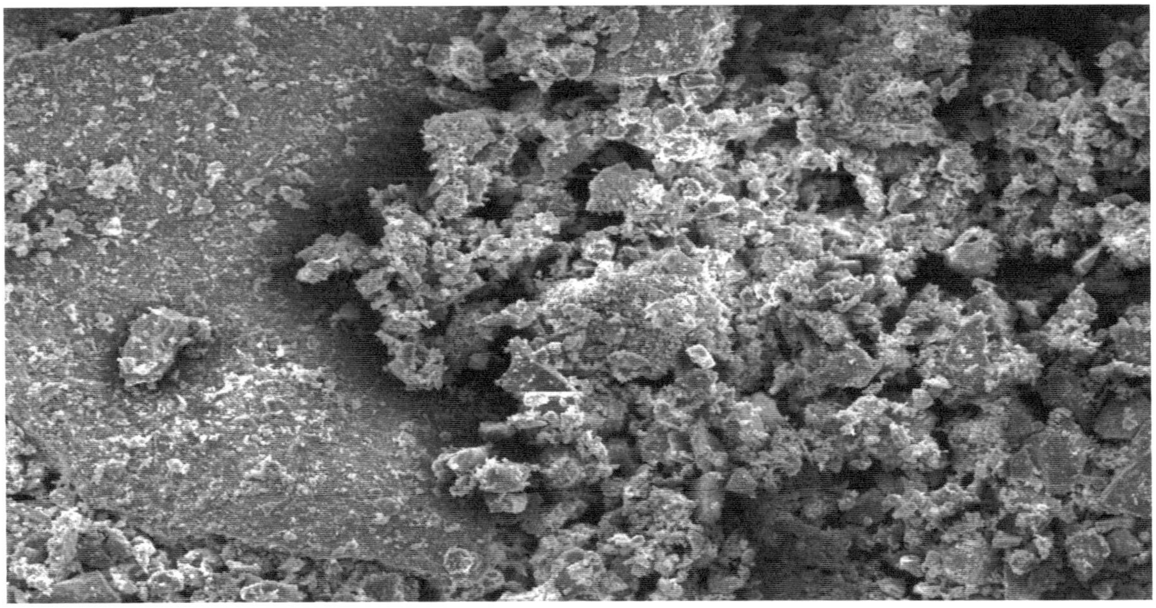

Figure 3: SEM analysis of the reacted calcite after 42 days

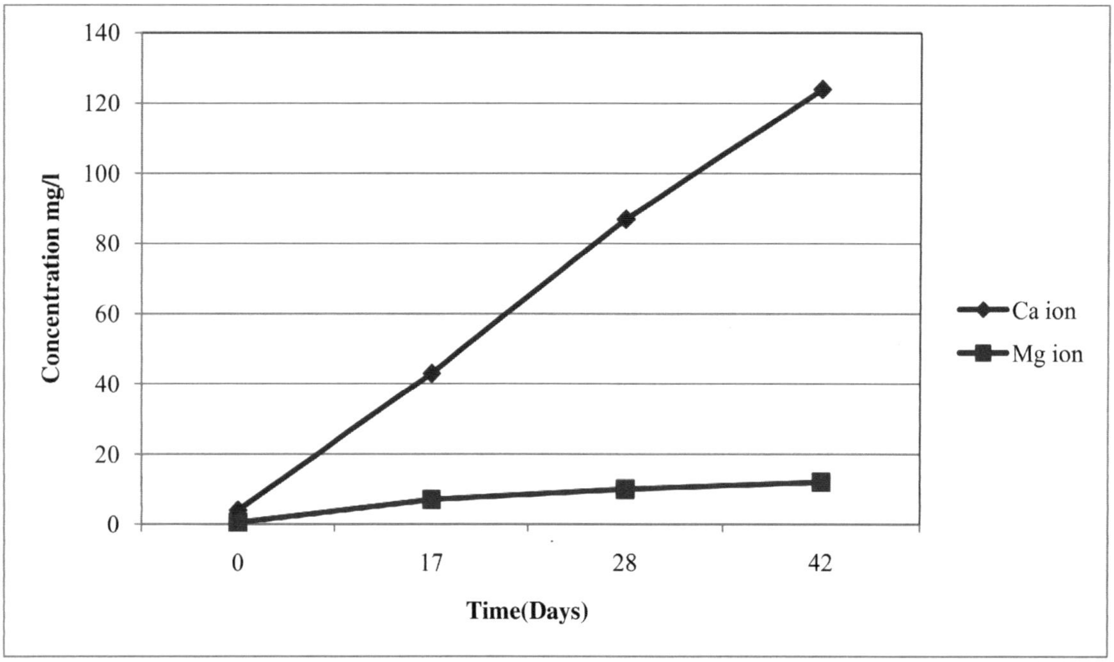

Figure 4: Principal changes in brine chemistry in the Limestone experiments

The primary ions tracked here are Ca ion and Mg ion (Figure 4). The Mg ion is derived from dolomite, which was present as traces in the initial mineral assemblage. The continuous increase in the concentrations of both the principal ions indicates dissolution of calcite and dolomite. The absence of any alumino silicate mineral to buffer the carbonic acid rules out the possibility of precipitation of any carbonates in a limestone aquifer. The primary modes of sequestration in these aquifers are structural trapping and ionic trapping.

Set B experiments with Sandstone

Initial XRD analysis of the sandstone revealed 44% calcium feldspar, 26.2 % Na feldspar and the rest potassium feldspar. During the experiments pressure was constant around 2000 psi after 17 days remained stable for the rest of the experiment. The pH increased from a value of 5.4 to about 5.9 after 28 days and 6.4 after 42 days. The principal reactions in this case can be formulated as follows.

(Eq 5) $2H^+ + CaAl_2Si_2O_8 + H_2O \leftrightarrow Ca^{2+} + Al_2Si_2O_5(OH)_4$

(Eq 6) $CaAl_2Si_2O_8 + H_2CO_3 + H_2O \leftrightarrow CaCO_3 + Al_2Si_2O_5(OH)_4$

(Eq 7) $NaAlSi_3O_8 + 3 H_2O \leftrightarrow NaAlSi_2O_6.H_2O + H_4SiO_4$

(Eq 8) $2KAlSi_3O_8 + 9H_2O + 2H^+ \leftrightarrow Al_2Si_2O_5(OH)_4 + 2 K^+ + 4H_4SiO_4$

Dissolution of feldspar buffers the brine (Eq 5), which turned acidic because of the formation of the carbonic acid. This leads to the next sequence of reactions, which is the carbonation of calcium feldspars (Eq 6), which leads to the precipitation of calcite and kolinite. At these temperatures and pressures albite undergoes phase change (Eq 7) to precipitate analcime which was detected in the XRD analysis. Dissolution of potassium feldspars (Eq 8) also leads to precipitation of kaolinite (Figure 7). The primary product of these dissolution reactions is silica. This leads to the brine becoming saturated and in some cases supersatured with silica. Hence when the samples are retrieved for analysis this silica undergoes heterogeneous deposition as amorphous silica on the surface of other host minerals, which was evident in the EDS analysis on all reacted samples. There was also evident precipitation of halite on these samples (Figure 5 and Figure 6). Halite precipitates when the sample is prepared for analysis, ie., drying overnight.

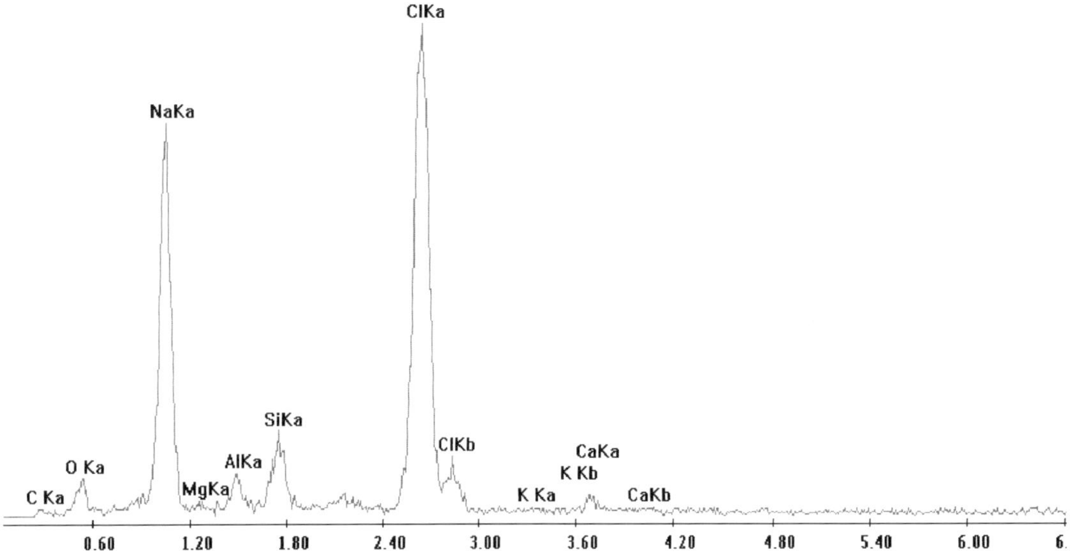

Figure 5: EDS analysis of halite

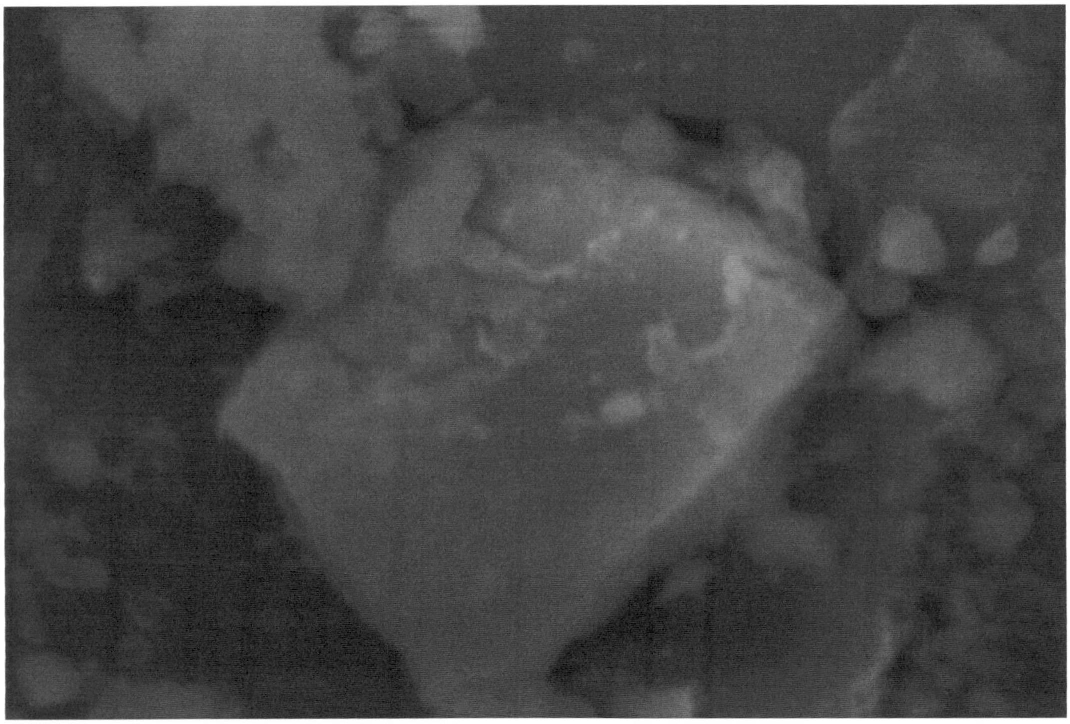

Figure 6: Halite chunk

Figure 7: Deposition of kaolinite on plagioclase feldspar

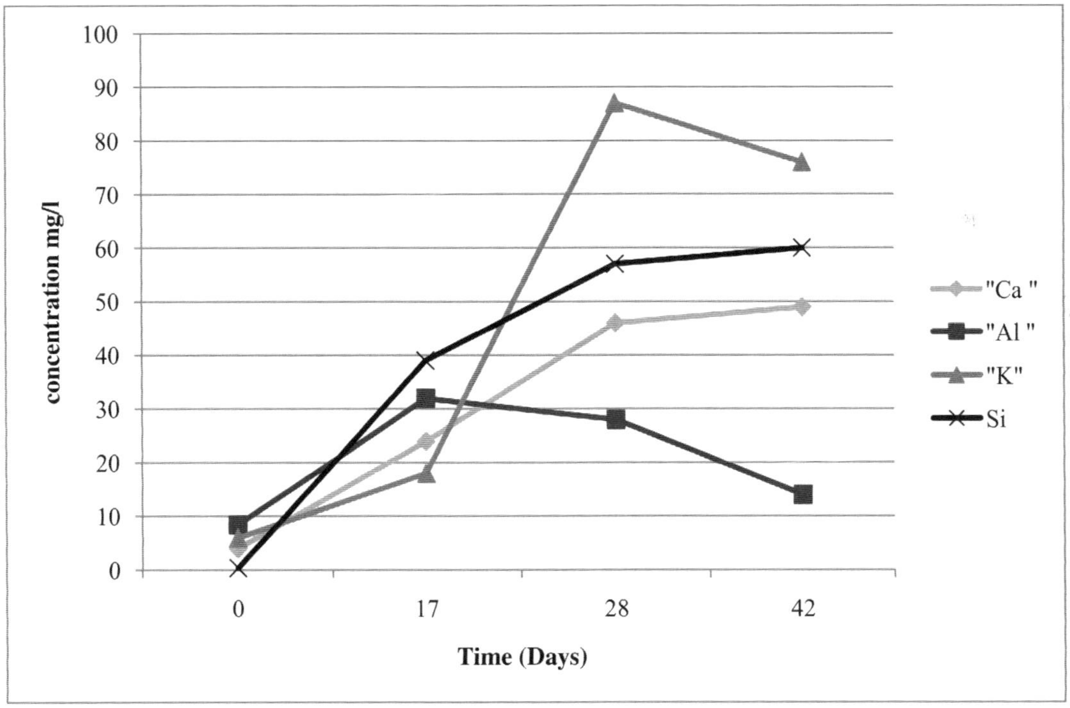

Figure 8: Principal changes in brine chemistry

Figure 8 shows the trend of Ca, K , Al and Si in the brine with plagioclase feldspar and microcline being the primary alumino silicates in the initial mineral matrix. No precipitation of carbonates was found in these experiments. The decrease in Al ion indicates its precipitation in solid phase, which is evident from the precipitation of kaolinite which is aluminosilicate hydroxide. Analcime was detected in XRD but its precipitation was not evident in the reacted samples.. Si concentration increases continuously in the samples because of the dissolution of the alumino silicate minerals.

Set C: Experiments with Peridotite

Peridotite is composed largely of the minerals Olivine [$(Mg,Fe)_2SiO_4$] and pyroxene [$(Ca,Mg,Fe)_2Si_2O_6$] which react with CO_2 and H_2O near the earth surface to form hydrous silicates (Serpentine), Iron oxides (magnetite) and carbonates (calcite, magnesite and dolomite). These reactions can be formulated as:

(Eq 9) $2Mg_2SiO_4 + Mg_2Si2O_6 + 4\,H_2O = 2Mg_3Si_2O_5(OH)_4$

(Mg-olivine) (Mg-pyroxene) (serpentine)

(Eq 10) $Mg_2SiO_4 + 2CO_2 = 2MgCO_3 + SiO_2$

(Mg-olivine) (Magnesite) (quartz)

(Eq 11) $Mg_2SiO_4 + CaMg_2Si_2O_6 + 2CO_2 + 2H_2O = Mg_3Si_2O_5(OH)_4 + CaCO_3 + MgCO_3$

(Mg-olivine) (Ca-Mg-pyroxene) (serpentine) (calcite) (magnesite)

Natural carbonation of peridotite by weathering and low temperature alteration is common. Evidence for natural, low-temperature hydration and carbonation of peridotite ca be found in springs and associated travertines inn catchments composed of peridotite and in outcrops of altered peridotite with abundant carbonate veins. High alkalinity, stable isotope ratios and formation of travertine and carbonate cemented conglomerates in springs indicate ongoing serpentenization involving meteoric water at low temperature. Ground water reacting with peridotite in near-surface, open systems forms water rich in

Mg^- and $HCO3^-$, which when on continuous reaction with peridotite leads to precipitation of abundant magnesite and dolomite as veins. The resulting waters become progressively richer in Ca- and OH-. These waters emerge near surface to mix with Mg-$HCO3^-$ waters or react with the atmosphere; they precipitate abundant calcite and dolomite in near surface veins, and carbonate cement in unconsolidated sediment and travertine. Enhanced natural processes may provide an important alternative to mineral carbonation.

Peridotite was obtained as big green crystalline granules from the deposits in the Samail Ophiolite, Sultanate of Oman. The distinctive green color was due to the presence of significant portions of Iron in the crystalline structure. Kinetics of these carbonation reactions is very slow unless olivine and serpentine reactants are ground to fine powder heated and hels at elevated pressures and temperature. Hence these granules were ground to very fine powder. A size distribution analysis revealed that major portion was within the range of 60-100 microns in size. 3 grams of peridotite was fed into a 40CC reactor along with 20CC of brine prepared by mixing 3grams of laboratory grade NaCl in 20CC of distulled water and brine was allowed to saturate the sample for about 2 days. CO_2 was then fed into the reactor at a pressure of 2000 psi and the temperature was maintained at 100C. The reactor was isolated and the experiment was carried out for 47 days. The pressure in the reactor first decreased and then stabilized around 1980 psi for the rest of the experiment with minimal fluctuations, which can be attributed to changes in ambient conditions.

Initial analyses of the sample revealed that Olivine was the major component in the sample. The sample was composed of 97.8 wt% olivine and 2.2% pyroxene (Figure 9 and Figure 10) . Hence we can conclude from the reactions mentioned above that carbonation of this sample will mainly yield magnesite and serpentine. Because of the presence of access CO_2 in the reaction setup, carbonation of peridotite will dominate hydration and hence magnesite precipitation should be omnipresent and formation of serpentine should be in trace amounts.

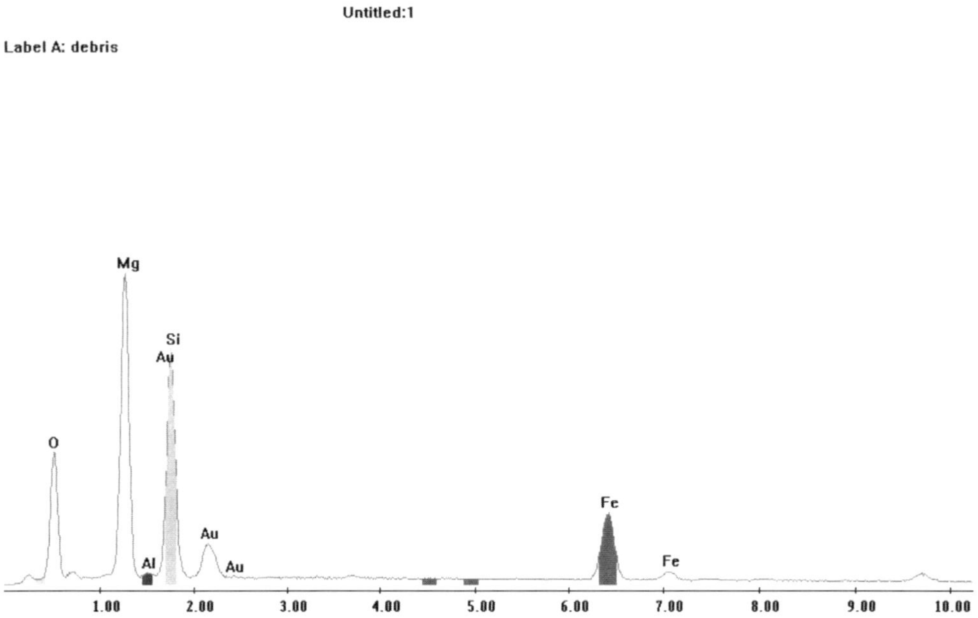

Figure 9: EDS analyses of the initial sample.

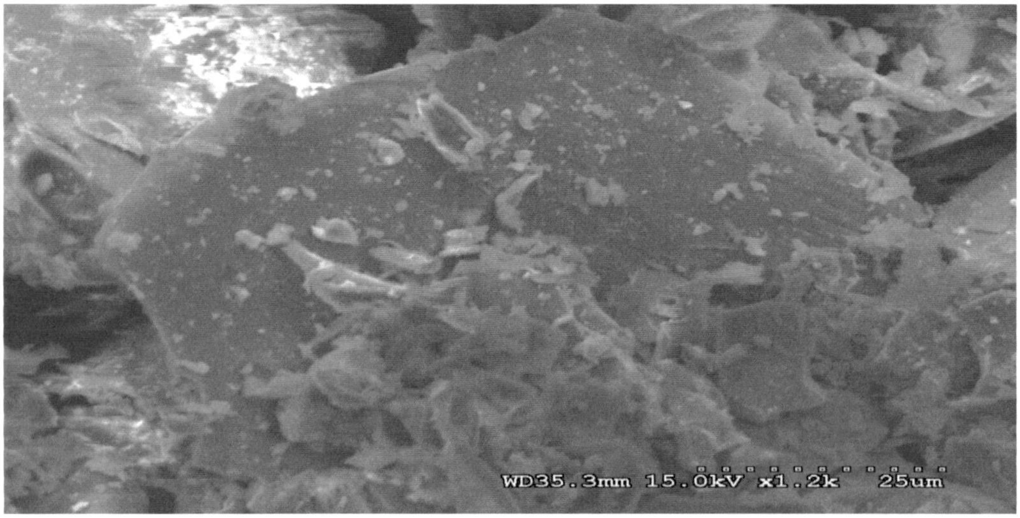

Figure 10 : SEM analyses of the initial rock

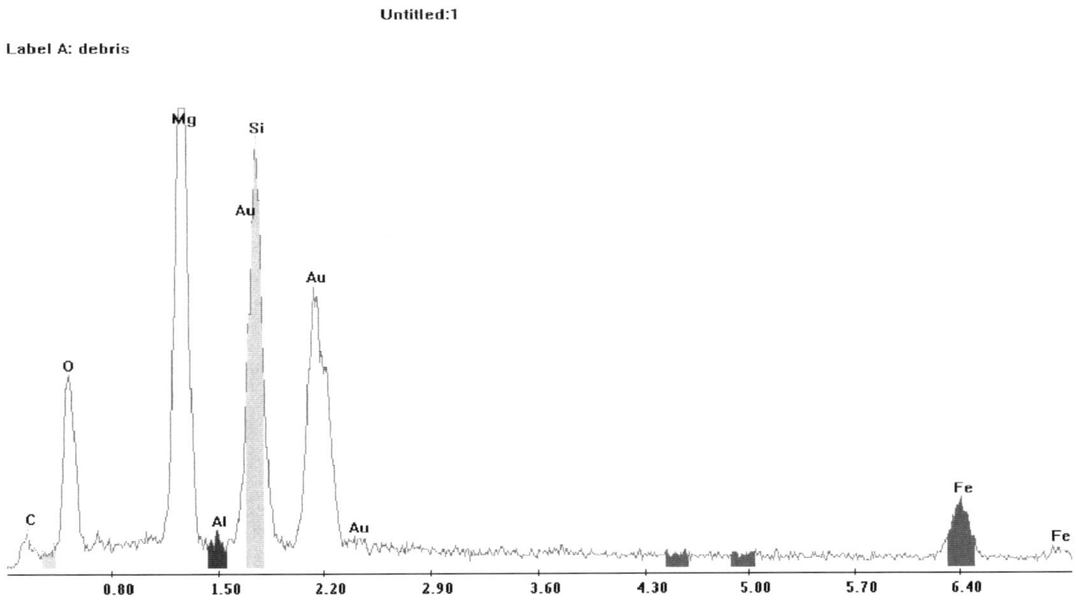

Figure 11: EDS analyses of the reacted rock.

The EDS analysis of the reacted rock (FIgure 11) indicates the presence of $MgCO_3$ in the sample. This occurs due to the carbonation of the peridotite to form magnesite and silica. This process is shown in the following reactions. The presence of silica is also revealed in the analysis. The SEM analysis of the product revealed the precipitation of magnesite as orthorhombic crystals (Figure 12) with pitted rough faces. Depending on the shape of the crystal it can be concluded that by the time the experiment the brine was under saturated with respect to magnesite. The dissolution of the alumino-silicate minerals produces silica. This silica precipitates out (Figure 13) mainly as amorphous silica on the products when the experiment is terminated and the temperature is decreased because the brine is supersaturated with respect to silica during the latter stages of the experiment.

Figure 12: Growth of magnesite crystal in the reacted sample

Figure 13: Trace amounts of silica mostly as fur on host peridotite

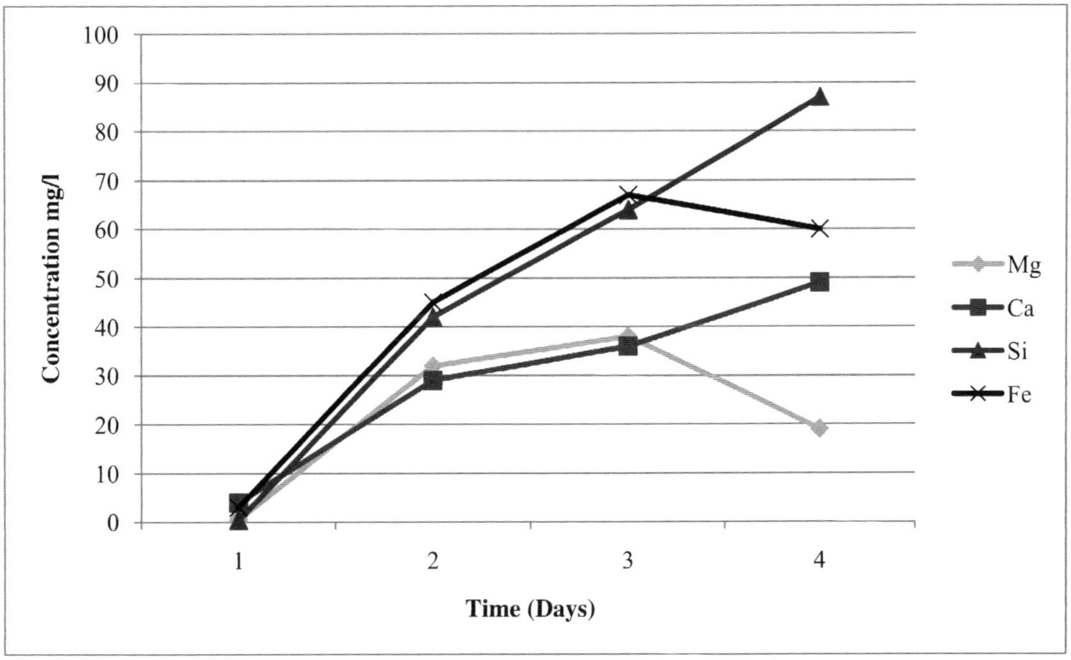

Figure 14: Principal changes in brine chemistry in Peridotite experiments

The initial increase in the primary ions Mg, Ca and Si (Figure 14) indicates the dissolution of the aluminosilicates, which leads to a decrease in the pH of the solution. This leads to the carbonation reactions (favored at higher pH) dominating the hydration reactions. The carbonation of peridotite leads to the precipitation of magnesite and siderite. The steep increase in the concentration of Si is a result of the dissolution of the silicate minerals. The reacted samples were characterized with the deposition of silica, which precipitated from the brine when the reactor was degassed.

SET D: Experiments with Arkose

All the participating minerals (constituents of arkose or dirty sand) (Table 3) are seen to be participating in the reaction. The major change is seen in the composition of dolomite and feldspars (microcline and andesine) with a small change in calcite and chlorite. The change in quartz composition was not expected and dissolution of quartz usually doesn't occur at pH 5. But as the pH increases and hits a pH of about 7 the H_4SiO_4 dissociates leading to the dissociation of quartz (Dreever 1988) . While dissolution patterns dominate the XRD analysis in the sample after 62 days precipitation patterns dominate after 134 days. Andesine, calcite and microcline are found precipitating while dolomite and quartz composition decreased in the final experiment. These results are in good correlation with the brine chemistry analysis and also the SEM/EDS analysis for the rock explained in the sections to follow.

Rock	Quartz	Andesine	Dolomite	Chlorite	Microcline	Calcite
Formula	SiO_2	$Na_xCa_yAlSi_2O_8$	$CaMg(CO_3)_2$	$(Fe, Mg, Al)_6(Si, Al)_4O_{10}(OH)_8$	$KAlSi_3O_8$	$CaCO_3$
Class	Silicates	Silicates	Carbonates	Silicates	Silicates	Carbonates
Group	Quartz	Feldspar	Dolomite	Chlorite	Feldspars	Calcite
Specific gravity	2.65	2.68-2.71	2.86	2.6-3.4	2.5	< 2.5
Hardness	7	6-6.5	3.5-4	2-3	6-6.5	3

Table 3: Composition of the arkose

Figures 15 and 16 show the SEM analysis of the reacted sample after 62 days and 134 days. The calcite in the initial samples serves as a standard to examine the precipitation and the dissolution patterns in the samples at different stages of the experiments. The precipitate from the experiment for 134 days was chosen because it was a reaction in which calcite precipitated as determined by the XRD. Figure 15 shows the pronounced dissolution patterns in the sample after 62 days. Carbon dioxide reacts with water to produce carbonic acid (Eq.12). And carbonic acid dissociates to bicarbonate ion in the equilibrium aqueous phase (Eq.13).

(Eq.12) $CO_2\,(g) \leftrightarrow CO_2\,(aq)$

(Eq.13) $H_2O + CO_2 \leftrightarrow H_2CO_3$

(Eq.14) $H_2CO_3 \leftrightarrow H^+ + HCO_3^-$

(Eq.15) $HCO_3^- \leftrightarrow H^+ + CO_3^{2-}$

Equation 12 is the dissolution of CO_2 in water, which is highly dependent on temperature pressure and salinity (Soong et al. 2002, Drever 1988). CO_2 solubility increases with increase in pressure, decreases with increase in temperature and salinity. CO_2 then reacts with water to from carbonic acid. This acid dissociates to liberate protons and bicarbonate ion. The sequential reactions in Eq.13 and 14 result in decreasing pH value (about 3.4). Dissociated hydrogen ion can dissolve calcite (mainly CaCO3) to produce calcium ion and bicarbonate (Eq. 16).

(Eq.16) $CaCO_3 + H^+ \rightarrow Ca^{2+} + HCO_3^-$

Dissolution occurs mainly by the formation of few deep etch pits and some shallow ones, almost at the same position in the initial dissolution of the surface. Dissolution occurs very fast which is followed by precipitation as seen in Figure 16 (Experiment carried out for 134 days).

The cations thus liberated (due to the increased acidity) react with the carbonate ion to precipitate secondary carbonates thus consuming CO_2 in the process.

(Eq.17) $Ca^{2+} + CO_3{}^{2-} \rightarrow CaCO_3$

Calcite formation can also be postulated by another mechanism: dissolution of the feldspars in the initial mineral matrix. For example arkosic sandstone as a carbonate reacts with hydrogen ion to precipitate calcium carbonate (Eq.18-20).

(Eq.18) $2H^+ + CaAl_2Si_2O_8 + H_2O \leftrightarrow Ca^{2+} + Al_2Si_2O_5(OH)_4$

(Eq.19) $Ca^{2+} + HCO_3^- \leftrightarrow CaCO_3 + H^+$

(Eq.20) $CaAl_2Si_2O_8 + H_2CO_3 + H_2O \leftrightarrow CaCO_3 + Al_2Si_2O_5(OH)_4$

The above reaction also shows kaolin deposition, which was evident in the reacted samples. This indicates that plagioclase feldspar is an active reactant in this geochemical repository. Calcite is seen growing as tightly packed polymorph of calcium carbonate. These layers and crystals of calcite are seen growing as amorphous mass intergrown with the starting minerals. These calcite crystals (Figure 17) are highly irregular in shape and show no consistency in size and shape. One calcite crystal had a width of about 10mm and most of the particles appear much smaller than this size. Hence it can be concluded not all precipitates were collected when the brine was filtered (with a filter paper of 8 mm retention capacity). Figure 18 shows deposition of a new phase analcime or kaolinite on the reacted surface.

The formation of analcime can be formulated by the transformation of albite (from the plagioclase feldspar)

(Eq.21) $NaAlSi_3O_8 + 3\ H_2O \leftrightarrow NaAlSi_2O_6.H_2O + H_4SiO_4$

Kaolinite formation from albite (from the plagioclase feldspar) can be formulated as

(Eq.22) $NaAlSi_3O_8 + CO_2 + 5.5H_2O \leftrightarrow Na^+ + HCO3^- + 2\ H_4SiO_4 + 0.5\ Al_2Si_2O_5(OH)_4$

Analcime deposits occur as large connected aggregates or as deposits on the surfaces of other minerals (Quartz). Kaolinite was not identified in the XRD analyses because it shares the same primary peak with one of the minerals in the initial assemblage, chlorite. But it was identified in the reacted assemblage. The growth was seen as aggregated units in the interstitial spaces between the primary minerals especially quartz.

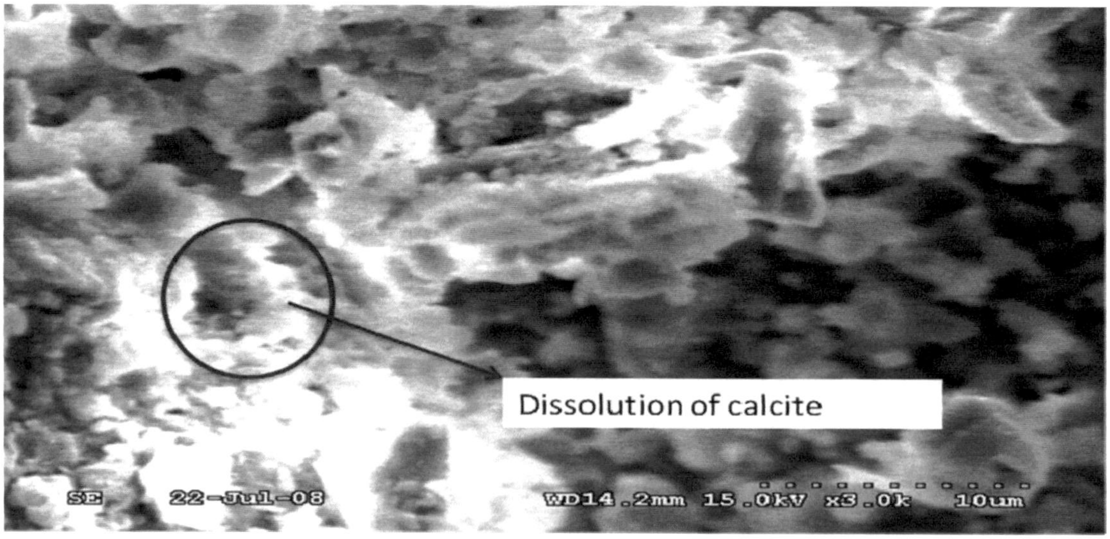

Figure 15: SEM image showing calcite in the sample showing dissolution after 62days.

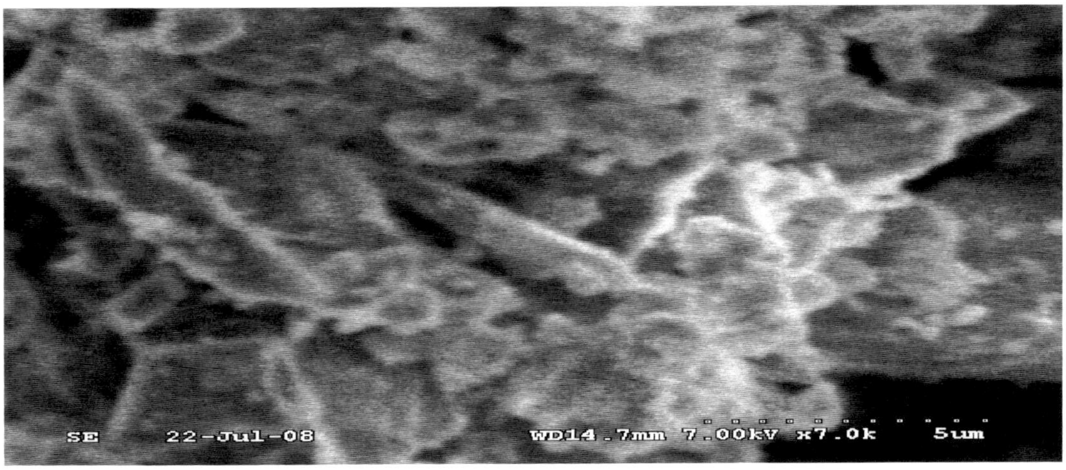

Figure 16: SEM image showing the growth of layers of Calcite after 134 days indicative of mineral precipitation reactions.

Figure 17: Calcite crystals growing in interstitial spaces between plagioclase feldspar.

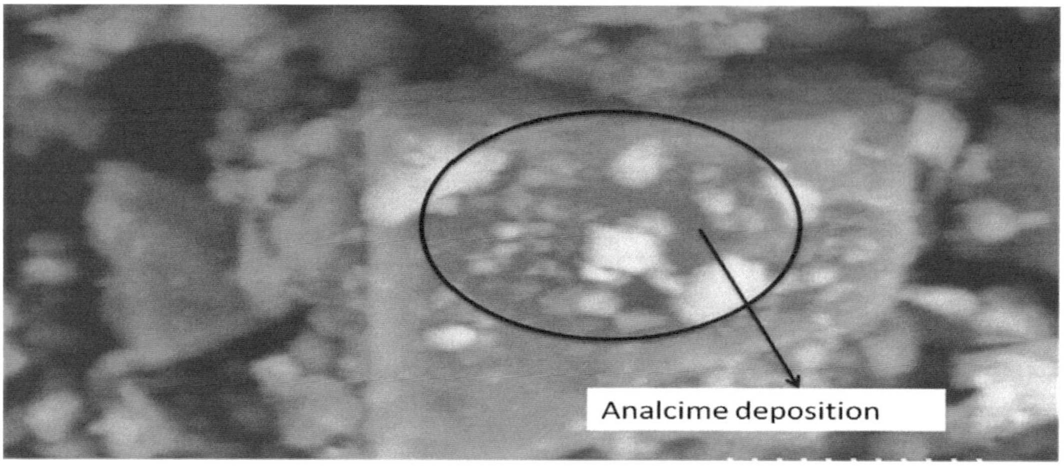

Figure 18: SEM image showing the growth of new phase probably analcime on the surface.

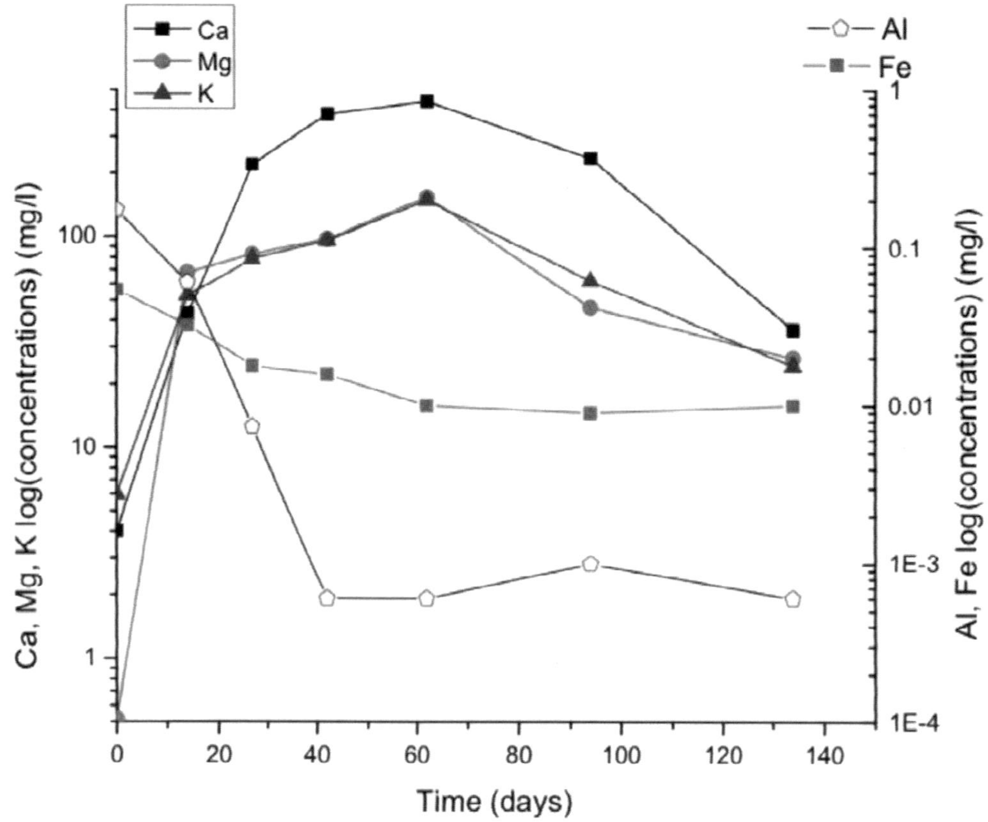

Figure 19: Changes in the concentration of principal ions

Ca ion concentration shot up by about 90% after 14 days and continued to increase for 42 days after which it decreased with the final concentration approximately 47% less than the highest concentration measured. The K ion concentration exhibited similar trend to that of Ca ion with the concentration shooting up abruptly and continuing to increase till 62 days after which it decreased. The increase in the Ca ion concentration can be attributed to the dissolution of the primary carbonate minerals calcite and dolomite liberating Ca ion into the brine. The K ion concentration increase was due to the dissolution of the microcline, which is potassium feldspar, and the silicate dissolution reactions are the fastest in a low pH geochemical system. The Mg ion concentration increased by approximately 52% at 27 days and continued to increase until 62 days after which it decreased with the final concentration 42% lower than the initial concentration measured at 27 days. This increase was mainly due to the dissolution of dolomite and chlorite. Iron (Fe) concentration decreased throughout the experiment and ended up approximately 27% lower than the initial concentration. The decrease in the concentrations of the Ca, Mg and K ions by 47%, 42% and 27% in the latter stages of the experiment are indication of the fact that new minerals with the primary composition of these ions are precipitating in the solid phase. The Si concentration followed a unique trend. It increased by 14% after 27 days then decreased slightly (2%) and then increased by about 32% till the termination of the experiment. Since the increase of 2% was within the experimental and analytical uncertainty, the concentration can be considered stable from 27 days to 42 days. The Si concentration increase can be attributed to the dissolution of feldspars, microcline and andesine, which are the most sensitive minerals to decrease in pH (Dreever, 1988). Hence the silicate dissolution dominates the geochemical reactions in the system. This is evident from the fact that there is a silica coating on all the samples analyzed in the SEM with a distinctive Si peak in the EDS analyses. This is due to the deposition of Si from the brine on to the rock when the solid sample is dried prior to the analysis.. The samples collected at the end of 62 days and 134 days were selected to represent these changes and Ca ion was chosen as the principal ion undergoing the change. At the end of 62 days and also through the first 62 days Ca ion concentration increased in the brine indicating the dissolution of primary carbonate minerals calcite and dolomite in the brine. It is clearly seen in the SEM analysis of the rock sample that is collected at the end of 62 days where calcite dissolution pattern dominates. Then, from 62 days through 134 days, the Ca ion concentration decreases, which implies that Ca bearing minerals should precipitate in the solid phase. This is clear from the SEM analysis of the sample collected at the end of 134 days where layers and crystals of calcite are seen growing as amorphous mass intergrown with the starting minerals (especially quartz).

The complex trend in the brine chemistry represents the geological complexity of the system. The presence of carbonates and feldspars favors the precipitation of carbonates, thus permanently sequestering carbon dioxide. The primary reactions are dissolution of carbonates, feldspars and carbonation of plagioclase feldspar and microcline. This set of experiments was characterized by precipitation of calcite, analcime and kaolin. These changes were complemented by the changes in brine chemistry.

Discussion

The carbonate system is highly pH dependant. The precipitation of thermodynamically stable carbonates depends on the pH of the brine in the geological repository. The pH inturn depends on the composition of the rock in the aquifer. This interdependency is very important and a proper understanding of the thermodynamics and kinetics of different reservoirs is quintessential to predict the ultimate fate of CO2 in the reservoir. The entire scenario boils down to the pH dependency of the carbonate system.

Reaction	Explanation	Equilibrium constant (25°C)
$CO_{2,g} + H_2O$ <====> $H_2CO_3^*$	Dissolution of CO_2	$pK_H = 1.5$ (at 25°C)
$H_2CO_3^*$ <====> $HCO_3^- + H^+$	First deprotonation stage	$pK_1 = 6.3$ (at 25°C)
HCO_3^- <====> $CO_3^{2-} + H^+$	Second deprotonation stage	$PK_2 = 10.3$ (at 25°C)

As we can see form Table 4, carbonate precipitation is favored at high pH. Due to the formation of carbonic acid, initially, the pH of the brine decreases. So the pH of the brine needs to increase to facilitate the precipitation of carbonates, This can only be achieved by buffering the brine. Hence the presence of alumino silicates (feldspars) or magnesium silicates (olivine or serpentine) in the mineral assemblage is necessary for carbonate precipitation. These minerals are very sensitive to changes in pH. They undergo instantaneous dissolution in acidic environments and hence buffer the brine. In addition to this, these feldspars and magnesium silicates also provide the necessary cations for the carbonate precipitation. The kinetic rates of dissolution of these minerals increase with an increase in temperature, but the dissociation constant of carbonic acid decreases with an increase in temperature, which means that the initially weaker carbonic acid will become increasingly weaker at higher temperatures. Hence there must be trade off while choosing an appropriate temperature for these reactions.

Each mineral behaves in a different manner in each assemblage. For example, let us consider calcite as an individual mineral and calcite in the arkose. The chemical composition of the mineral is the same, but the mineral assemblage it is placed within, plays the governing role in its ultimate fate. In the case of carbonate reservoir, calcite undergoes dissolution releasing Ca ion and CO32- or HCO3- ion depending on the stage of deprotonation and pH. Thus it traps CO2 in ionic phase or through structural trapping, Hence addition of CO2 to a carbonate reservoir will increase acidity and sissolve, not precipitate carbonate minerals and CO2 will remain as dissolved aqueous species or free gas. In a arkose reservoir, the initial reactions of calcite are the same. The parallel reactions that take place are the dissolution of feldspars and chlorite. These reactions trigger a sequence of carbonation reactions. The cations are added to the brine by the dissolution of feldspars combine with the carbonate or the bicarbonate ions and precipitate, thus sequestering CO2. Hence the presence of feldspars or magnesium silicate minerals is necessary for mineral sequestration of CO2 through carbonation.

In sandstone reservoirs the relatively slow carbonation rates when compared to the rapid dissolution of the silicate minerals underplays its significance. The carbonation of peridotite yields magnesite and sequesters CO2 as carbonates. But the relative abundance of dirty sandstones to peridotite makes it an apt choice as the initial mineral assemblage.

References

Bachu, S., 2002. Sequestration of Carbon dioxide in geological media in response to climate change: road map for site selection using the transform of the geological space into the CO_2 phase space. Energy Convers. Manage 43 (1), 87-102.

Bacon , D, H., Sass. M .B., Bhargava. M., Sminchak. J.,, and . Gupta. N., 2009. Reactive Transport Modeling of CO_2 and SO_2 Injection into Deep Saline Formations and Their Effect on the Hydraulic Properties of Host Rocks. Energy Procedia (1) 3283–3290.

Bachu, S.; Gunter, W.D; Perkins, E.H., 1994. Aquifer disposal of CO_2 : hydrodynamic and mineral trapping. Energy Converse. Manage. 35 (4), 269-279.

Bateman, K.; Turner, G., 2005. Large scale column experiment: Study of carbon dioxide, pore water, rock reactions and model test case. Oil and Gas Tech 60, 161–175.

Bergman, P.D.; Winter, E.M., 1995. Disposal of CO_2 in aquifers in the U.S. Energy Conver. Manage 36, 523-526

Climate Change, 2007 An assessment of the Intergovernmental panel on climate change. http://www.ipcc.ch/pdf/assessment-report/ar4/syr/ar4_syr.pdf

Druckenmiller, M. L.; Maroto-Valer, M. M.; Hill, M., 2005. Investigation of carbon sequestration via induced calcite formation in natural gas well brine. Energy and Fuels.

Dreever, J.I., 1988. The Geochemistry of Natural Waters: Surface and Groundwater Environments, Prentice Hall, INC.

Gunter, W.D.; Perkins, E.H.; Hutcheon, I. 2000. Aquifer disposal of acid gases: modeling of water rock reactions for trapping of acid wastes. Appl. GeoChem 15 (8), 1085-1095.

Gunter, W.D.; Perkins, E.H.; McCann, T.J., 1993. Aquifer disposal of CO_2 rich gases: reaction design for added capacity. Energy Converse. Manage. 34, 941-948.

Jacquemet, N.; Pironon, N.; Marc, J. S., 2008. Mineralogical changes of a well cement in various H_2S-CO_2 (brine) fluids at high pressure and temperature. Environ. Sci. Technol 42, 282–288.

Kaszuba, J. P. ; Janecky, D. R. ; Snow, M. G., 2005. Experimental evaluation of mixed fluid reactions between supercritical carbon dioxide and brine: Relevance to the integrity of a geologic carbon repository. Chem Geol 217, 277-293.

Kaszuba, J. P. ; Janecky, D. R. ; Snow, M. G., 2003. Carbon dioxide reaction processes in a model brine aquifer at 2000 C and 200 bar: Implications for geologic sequestration of carbon. Appl Geo. Chem 18, 1065–1080.

Lindeberg, E., 1997. Escape of Carbon Dioxide from aquifers. Energy converse. Manage 38, 238–240.

Moore, J.; Adams, M.; Allis, R.; Lutz, S.; Rauzi, S. 2005. Mineralogical and geochemical consequences of long term presence of CO_2 in natural reserviors: An example from the Springerville-St.Johns field, Arizona and New Mexico, U.S.A. Chem Geol 217, 365–385.

Moore, J.; Christensen, W.B; Allis, R.; Browne, L.R.; Lutz, J.; 2004.The Minrealogical consequences and the results of descending acid sulphates water: An example from the Karaha – Telaga-Bodas Geothermal system, Indonesia. Can.Min vol 42. pp 1483-1499.

Rosenbauer, R. J.; Koksalan, T; Palandri, J.L, 2005. Experimental investigation of CO_2 -brine-rock interactions at elevated temperature and pressure : Implications for CO_2 sequestration in deep-saline aquifers," Fuel Process. Tech 86, 1581–1597.

Span, R.; Wegner, W., 1996. A new equation of state for carbon dioxide covering the fluid region from the triple point temperature to 1100K at pressures up to 800 MPa. J. Phys. Chem. Ref. Data 25 (6), 1509-1596.

Soong, Y.; Goodman, A. L.; Hedges, S. W.; Jones, J. R.; Harrison, D. K., 2002. CO_2 Sequestration and importance of pH. Am. Chem. Soc 47, 43–48.

Seyfried Jr.; Janecky, W.E.; Berndt, D.R., 1987. Rocking autoclaves for hydrothermal experiments, II. The flexible reaction cell system. In: Ulmer, G.C.; Barnes, H.L.

U.S. Department of Energy, 2007, Carbon Sequestration Atlas of the United States, National Energy Technology Laboratories.

U.S. Department of Energy, 2008, Carbon Sequestration Atlas of the United States, Version 2, National Energy Technology Laboratories.

U.S. Department of Energy, 2009, Storage of Captured Carbon Dioxide on Federal Lands, National Energy Technology Laboratories, DOE/NETL-2009/1358.

(389c) Risk Assessment in CO_2 Sequestration

Yan Zhang, Nick Sahinidis and Panagiotis Vouzis, Chemical Engineering, Carnegie Mellon University, Pittsburgh, PA

A main concern for any sequestration system is whether it may leak CO_2 over a long-term time horizon consisting of hundreds or thousands of years. The outcome depends on the competition between sequestration and leakage processes. Leakages after sequestration in an abandoned reservoir, for instance, may occur from failure of the manmade cement seal, or through naturally existing fractures in the caprock. On the other hand, the CO_2 source pool may dissolve into formation water, migrate to water of surrounding formations, or react and mineralize. Modeling all these processes from first principles involves a number of parameters, such as various permeability and porosity values, that are not known with certainty.

To deal with uncertainty in model parameters, we develop a Monte Carlo simulator with analytical equations for geologic sequestration based on the CQUESTRA model [1]. To assess the uncertain long-term fate of CO_2, we carried out multiple simulations with parameter values drawn from probability distributions based on data from the literature. The proposed methodology was also parallelized on a Graphics Processing Unit, where 100 times speedup was achieved compared to a four-threaded CPU implementation. Computational results will be presented for several failure scenarios, suggesting that most of the CO_2 is safely trapped underground. Sensitivity plots further suggest that the leakage percentage is most sensitive to the permeability of the target reservoir.

Reference:

1. D. M. LeNeveu, "CQUESTRA, A Risk and Performance Assessment Code for Geological Sequestration of Carbon Dioxide," Energy Conversion and Management, 49(1), 32– 46, 2008.

See more of this Session: Carbon Dioxide Injection in Underground Formations
See more of this Group/Topical: Topical D: Chemical Engineering in Oil and Gas Production and Other Complex Subsurface Processes

(389d) Thermodynamic and Kinetic Studies of Mineral Trapping of Carbon In Geologic Formations

Greeshma Gadikota, Chemical Engineering, Columbia University, New York, NY and Ah-Hyung Alissa Park, Earth and Environmental Engineering and Chemical Engineering, Columbia University, New York, NY

The ever rising emission of anthropogenic CO2 has been one of the critical challenges faced by the global community. One of the most stable and long-term solutions for storing CO2 is via carbon mineralization where minerals containing metal oxides of Ca or Mg are reacted with CO2 to produce thermodynamically stable Ca- and Mg-carbonates that are insoluble in water. Carbon mineralization can be carried out ex-situ or in-situ. The kinetic of in-situ mineral trapping via carbonation is naturally very slow. However, ex-situ carbon mineralization involves mining and mineral processing, which can be quite energy-intensive. Thus, this work focuses on the chemical enhancement of in-situ mineral carbonation and the effect of mineral trapping on the stability of injected CO2 in geologic formations. Various silicate minerals such as olivine (Mg2SiO4), labradorite ((Ca,Na)(Al,Si)4O8) and basalt (mixture of silicate minerals) are selected for thermodynamic and kinetic studies. The kinetics of mineral dissolution and subsequent carbonation steps are first independently studied, and both Visual MINTEQ and Geochemist's Workbench are used for thermodynamic calculations. Particularly, the effects of ionic strength and chemical additives on the mineral dissolution, which has been known as a rate limiting step, are investigated using both a differential bed reactor and a batch reactor. The reaction parameters including temperature, pressure, reactive surface area, and pH are also studied. These results will provide important kinetic data to be used in the reaction path model for predicting the fate of injected CO2 in geologic formations.

See more of this Session: Carbon Dioxide Injection in Underground Formations
See more of this Group/Topical: Topical D: Chemical Engineering in Oil and Gas Production and Other Complex Subsurface Processes

(389e) A FULLY COUPLED GEOCHEMICAL COMPOSITIONAL SIMULATOR for Modeling CO2 Sequestration in Formations with Discrete Fault/ Fracture Networks

Milind D. Deo and **Zhiqiang Gu**, Chemical Engineering, University of Utah, Salt Lake City, UT

A geochemical coupled compositional model is developed to describe the complex processes during CO2 sequestration. The model is built on a modularized framework that allows coupling different discretization schemes with simulations of different physical processes. The modular approach allows the use of different spatial discretization methods including control-volume finite element method (CVFEM), standard finite difference method (SFD) and a generalized finite volume method (FVM). The coupled geochemical compositional model is capable of handling multi-phase and multi-component reservoir problems. The reaction option available in the model is able to handle complex geochemical systems. A new solution method is implemented to make the model more suitable to solve problems in complex geologic formations with complex geochemistry models. Natural variables are chosen as primary variables for their simplicity. Variable substitution and a novel equation alignment technique are required to correctly handle phase change, which is very common in the different phases of the CO2 sequestration process. For maximum numerical stability, a fully implicit scheme and an analytical Jacobian matrix are used, and all equations are solved simultaneously in the simulator. Gaussian elimination is applied twice before the solution of the linear equation system, first to eliminate equilibrium reaction rates and then to reduce the size of the linear system. The model is verified using analytical solutions and benchmarks. It is shown that a DFN representation is essential in correctly modeling the rapid and oftentimes uneven migration of CO2 to the seal. The DFN approach also facilitates modeling critical aspects of fault reactivity (dissolution) which may lead to seal breaches. A novel parallelization scheme for domains with complex fracture systems is also described. The CVFEM-based method is used for the first time to analyze the impact fault/fracture networks may have on the integrity of storage in the CO2 sequestration process.

See more of this Session: Carbon Dioxide Injection in Underground Formations
See more of this Group/Topical: Topical D: Chemical Engineering in Oil and Gas Production and Other Complex Subsurface Processes

Effects of wettability of rock-CO_2-water systems on the migration and distribution of geologically sequestered supercritical CO_2

Chuan Lu, Weon Shik Han, Brian McPherson
Energy & Geoscience Institute, University of Utah

Peter Lichtner
Computational Earth Sciences (EES-16), Los Alamos National Laboratory

Abstract

CO_2 sequestration (capture, injection, and long-term storage) in various geologic media including depleted oil reservoirs, saline aquifers, and oceanic sediments is being considered as a possible solution to reduce greenhouse gas emissions. Residual or capillary trapping is one of the most important mechanisms to impede the migration of supercritical (SC) CO_2 and reduce leakage risks. Capillary pressure and relative permeability are key factors in sequestration, and they are strongly affected by the interfacial properties between SC CO_2, saline fluids and formation rocks. Interfacial properties are usually described by interfacial tension and contact angle. Preliminary analyses regarding the effects of wettability and entry pressure on the capacity and sealing efficiency have been presented in the literature. A few experimental observations illustrate that interfacial properties are functions of pressure, temperature, salinity and pH of formation fluids, as well as the formation rock mineralogy. The injection of SC CO_2 can alter the in-situ conditions of P, T, pH, etc. in target formations, and consequentially change the interfacial tension and contact angle.

In this study, numerical simulations on idealized formations are conducted to investigate the effects of SC CO_2 injection induced interfacial property alteration on the capillary pressure and relative permeability, and consequently on the extent of SC CO_2 migration and storage capacity. Simulations and analyses in this study are mainly concentrated on three effects of SC CO_2 migration: 1) reservoir rock formation wettability; 2) capillary pressure variation; and 3) variations in relative permeability. To date, experimental measurements in the literature have been focused mostly on the variation of interfacial tension and contact angle. These effects are investigated in the research presented here by scaling the capillary pressure, calculated from the van Genuchten model, with parameters estimated from in-situ conditions (P, T, pH and salinity) using correlations determined from previous studies. By contrast, experimental investigations of the dependence of relative permeability on wettability are relatively sparse in the literature. In order to facilitate our investigation, the modified Burdine model is adapted, in which the effects of contact angle are explicitly described. It is found that the SC phase irreducible saturation increases as the SC CO_2 wettability of the reservoir formation rock is elevated, thus enhancing storage efficiency.

Introduction

CO_2 sequestration (capture, injection, and long term storage) in various geologic media (such as depleted oil reservoirs, saline aquifers, oceanic sediments) is being considered as a possible solution to reduce green house gas emissions. This approach has been widely discussed in the literature during the last decade [1]. Field-scale CO_2 sequestration experiments are currently being conducted in various locations and geologic environments around the globe. For example, large scale CO_2 disposal in deep saline aquifers has taken place in the Norwegian sector of the North Sea (Sleipner) since 1996 [2].

It is widely accepted that the mechanisms to prevent sequestered CO_2 from escaping to the surface include hydrodynamic (mobile), residual, solubility and mineral trapping [3, 4, 5]. "Hydrodynamic trapping" is the storage of supercritical (SC) CO_2 as a mobile phase below a low permeable formation or cap rock. "Residual trapping" is the storage of SC CO_2 in pores due to capillary pressure. The time scale of both hydrodynamic and residual trapping is generally considered to be on the order of hundreds of years and they are the most important trapping mechanism in the early stages of CO_2 sequestration. The characteristics and efficiency of residual trapping are strongly dependent on the relation of saturation, capillary pressure and relative permeability (s-p_c-k_r). These relations are in turn dependent on the in situ properties of reservoirs, e.g. fluids and rocks, pressure, temperature and salinity etc. and their interactions, such as interfacial tension and wettability.

Currently, numerical simulations of geological CO_2 storage assume reservoir rocks are water wetted. However, some minerals are oil or SC CO_2 wetted. Polar compounds or asphaltenes deposited from the crude oil onto mineral surfaces cause the oil-wet condition. Similar compounds in oil-base mud also can cause a previously water-wet rock to become partially or totally oil-wet. The injection of SC CO_2 in formations can alter fluid-rock interactions, hence change the interfacial tension and wettability [6, 7]. Though there is no experimental evidence showing that the injection of SC CO_2 can change a mineral from water wetted to SC CO_2/oil wetted, the contact angle between SC CO_2 and aqueous phases does depend on the system pressure [7]. This behavior may be caused by the dissolution of SC CO_2 in water and the consequential modification of interfacial properties. As these interactions have not been widely investigated, this work attempts to investigate these phenomena.

Model description

In order to investigate the effects of varying s-P_c-K_r relations on plume migration and distribution, certain model conditions should be satisfied. The model should: 1) be able to describe water, SC CO_2 and fractional wetted system; 2) capture hysteresis effects to evaluate residual trapping and; 3) easily be modified to dynamically vary the surface tension and wettability during simulation processes. Based on these considerations, the s-P_c-K_r relations adapted in this study are very similar to the ones used by [8]. In this model, hysteresis wettability effects were taken into account by distinguishing between immobile and entrapped residual saturation. The concept of immobile saturation S_{ai} (a=o,w, the phase index) is only applicable when applied to the wetting phase; it is caused by a wetting thin film coating the solid surface. On the other hand, the concept of entrapped saturation (S_{at}) is only applicable to the non-wetting phase; it occurs in the larger portion of the pore space as a wetting phase intrudes due to capillary imbibition.

In this model, the effects of wettability on the S-P_c relation is evaluated by scaling the perfectly wetted capillary pressure with the macroscopic contact angle with Equation 1 to obtain the wettability dependent P_c:

$$P_C^*(\overline{S}_W^{app}) = \cos(\phi_{SOW}) P_C(\overline{S}_W^{app}) \quad (1)$$

where the P_c is the capillary pressure evaluated with the van Genuchten model

$$P_C = \frac{1}{\alpha}[\overline{S}_W^{app^{-\frac{1}{m}}} - 1]^{\frac{1}{n}} \quad (2)$$

where α is the model parameter and m is assigned as 1-2/n in order to simplify the implementation of relative permeability [8]. ϕ_{sow} is the contact angle. Figure 1 shows the capillary pressure of ϕ_{sow} =0 and 180, respectively.

The variable s^{app}_w is the apparent water saturation, defined as the summation of mobile and entrapped saturations; detailed discussion can be found in [8]. With those definitions, based on the modified Burdine model, Bradford deduced the corresponding relative permeability model:

$$k_{rw}(\overline{S}_W^{app}) = (\overline{S}_W^{eff} - \overline{S}_{Wt}^{eff})^2 \frac{[1 - f(\phi_{SOW})]\int_0^{\overline{S}_W^{app}} R(S)^2 dS + f(\phi_{SOW})\int_{1-\overline{S}_W^{app}}^1 R(S)^2 dS}{\int_0^1 R(S)^2 dS}, \quad (3)$$

and

$$k_{ro}(\overline{S}_o^{app}) = (\overline{S}_o^{eff} - \overline{S}_{ot}^{eff})^2 \frac{f(\phi_{SOW})\int_0^{\overline{S}_o^{app}} R(S)^2 dS + [1 - f(\phi_{SOW})]\int_{1-\overline{S}_o^{app}}^1 R(S)^2 dS}{\int_0^1 R(S)^2 dS}, \quad (4)$$

for wetting and non-wetting phases correspondingly. Within our applications, the phase subscript w and o imply the aqueous and SC/gaseous CO_2 phase respectively.

In equation (3) and (4), $R(S_w) = P_c(S_w)/2\sigma_w$. If the capillary pressure is evaluated with the van Genuchten model (Equation (2)), and with m=1-2/n, the integral $\int_a^b R(S)^2 dS$ is given by [9] as

$$\int_a^b R(S)^2 dS = [1 - a^{1/m}]^m - [1 - b^{1/m}]^m. \quad (5)$$

Instead of applying the concept that wettability is a function of saturation as [8], we assume that wettability is a simple linear function (Equation 6) of dissolved CO_2 concentration

$$\cos(\phi) = 1 - \frac{X_{CO2}}{X_{\max}} \quad (6)$$

where X_{max} is the CO_2 saturation mole fraction at a given pressure and temperature. This formula allows the cos(ϕ) to vary between 0 to 1, i.e. the rock wettability can change from water wetted to equally water/CO_2 wetted. This assumption is hypothetical and very simplified. When more detailed information appears in the literature, it will be modified. All numerical simulations are conducted with simulator PFLOTRAN [10].

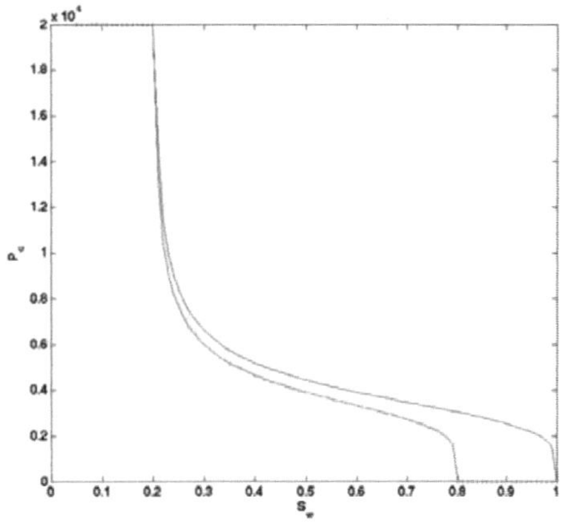

a) P_c with cos(φ)=1

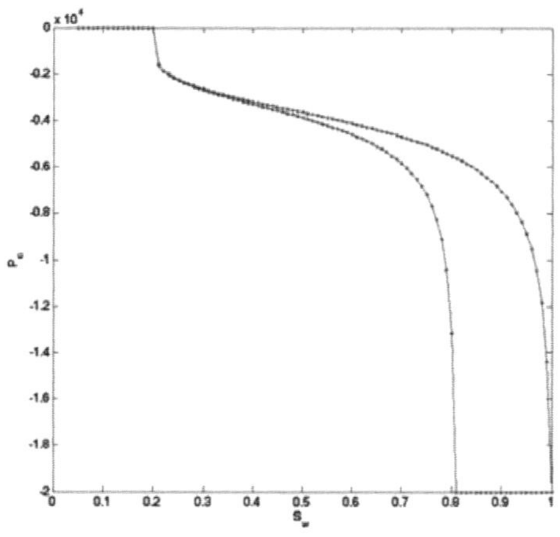

b) P_c with cos(φ)=-1

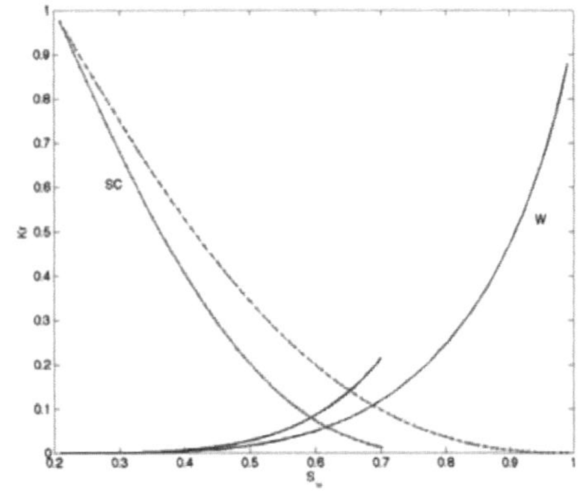

c) kr with cos(φ)=1

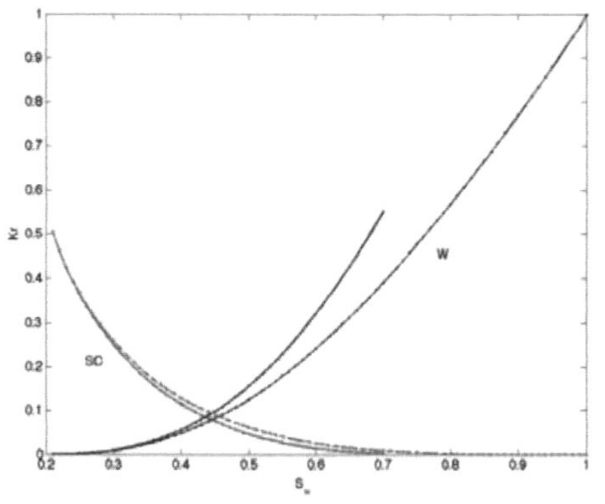

d) kr with cos(φ)=-1

Figure 1. The s-P_c-K_r curve investigated. (S_{ir}=0.2 for both wetting and non-wetting phase).

Numerical simulations and discussions

1D simulations

Simple 1D simulations can illustrate the effects of wettability on the features of buoyancy driven migration of SC CO_2 in geologic formations. In our simulations, the size of the simulated 1D vertical domain is 1*1*128 m, and discretrized with a 1*1*128 evenly spaced grid. A uniform permeability of 10 mD and porosity of 0.15 are assigned throughout the model domain. No flow BC at the bottom boundary and constant pressure (20MPa) and saturation (S_w=1) at top the boundary were applied. At the beginning of the simulation, the SC CO_2 plume with a saturation S_g=0.7 resides at depth from 65-96m. The media is fully saturated with water otherwise.

Effects on migration speed

Six scenarios were investigated, in which the formations are assumed to be water wetted, SC CO_2 wetted or varying with dissolved CO_2 concentration, with and without capillary pressure, as summarized in Table 1.

Table 1 Configurations of 6 cases studied to investigate the migration speed

Case	$cos(\phi)$	P_c(0/1)
A	0	0
B	1	0
C	-1	0
D	1	1
E	-1	1
F	0--1	1

The migration speed can be characterized with the migration of the SC phase mass center, evaluated by moments of the SC phase. The 2nd order moments M_{ij} (i,j=1,2) in 2D system is defined by (11)

$$M_{ij} = \int_{-\infty}^{\infty}\int_{-\infty}^{\infty} \phi \rho s_{CO_2} z dx dz \qquad (7)$$

with the mass center defined as $Z_c=M_{01}/M_{00}$.

Figure 2 shows the evolution of the mass center of the SC plume for 6 scenarios. After 25 years, the migration speed of the mass center of Cases C and E are similar to each other and are the fastest among the 6 cases. The speed in Cases B and D are similar to each other and are the slowest among the 6 cases. Prior to year 25, the migration speeds are very closely matched. Bouyancy driven migration stops within about 25 years in water wetted formations, and lasts for about 125 years in SC CO_2 wetted formations. These migration effects are essentially caused by hysteresis, rather than the driving force and magnitude of relative permeability. These aspects are explained in the next section. These results also indicate that the effect of capillary pressure does not strongly influence the migration speed,which may be caused by the capillary pressure model used (van Genuchten type) which has zero entry pressure.

Another behavior to be noted is that the migration speed of the mass center does not determine the speed of the displacement front, because the distribution patterns of the SC phase plume are different in rocks with different wettabilities. This phenomena is discussed in the next section.

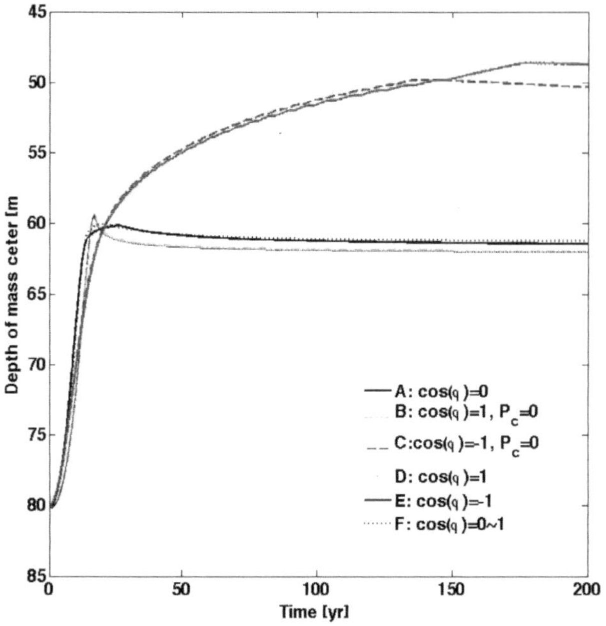

Figure 2. SC CO_2 plume migration speed in the 1D vertical model.

Effects on plume distribution

Four scenarios are simulated to investigate the effects of wettability on the plume distribution. The domain size, grid setup, boundary and initial conditions are identical to those used in the previous section. The rock wettability is set according to Table 2. These configurations form layered composite systems. The prefix S and F specify seal and formation rock, respectively, though they are assigned the same permeability values. The prefix W and C specify water and SC CO_2 wetting, respectively.

Table 2 Rock wettability configuration to investigate the effects on plume distribution

Case	Grid region				
	1-16 (F)	17-32 (S)	33-48 (F)	48-64 (S)	65-128 (F)
1 (SW,FW)	W	W	W	W	W
2 (SC, FC)	C	C	C	C	C
3 (SW, FC)	C	W	C	W	C
4 (SC, FW)	W	C	W	C	W

Figure 3 shows the SC phase saturation at 100 years. Excluding the deepest part of the model (Grid node 65-128) where the initial condition still has effects, the general features include:

1. SC phase saturation is higher in SC CO2 wetted regions than that in water wetted regions. Consequently, the SC phase plume "prefers" to stay in SC CO2 wetted regions.
2. SC phase saturation in the deepest part of the formation is close to 0.3 if this region is water wetted. This value is equal to the trapped SC phase saturation as predicted by S-kr relations shown in Figure 1 on the 1st order imbibition curve.
3. The lowest SC phase saturation in the SC CO2 wetted region is close to 0.14, which is consistent with the irreducible SC phase saturation predicted by $S_{io}=S_{ro}^{*} \times S_{o,max}$, where S_{ro}* is the "reference" irreducible oil saturation and So,max is the historic maximum oil saturation [8].
4. The effect of capillary pressure causes the SC phase to accumulate below the water wetted layers.

The lower irreducible SC phase saturation $S_{i,SC}$ allows more CO_2 to stay in the mobile state resulting in buoyancy driven upward migration to last longer, as shown in the last section.

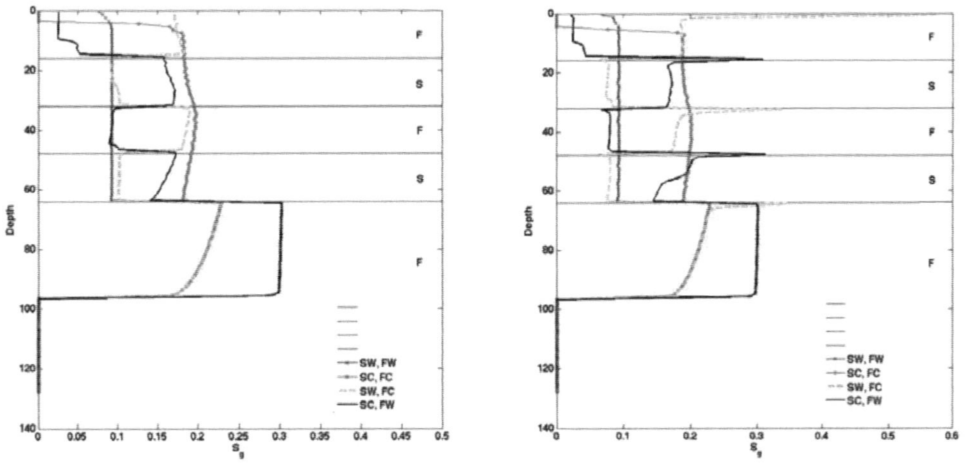

a) P_c=0 b) P_c (Equation 1)

Figure 3. SC phase saturation distribution in 1D layered systems.

2D simulations

The 2D simulations are conducted on a domain size of 4000x192m, discretized with 40x15m on an evenly spaced 40x128 grid. The same top and bottom BC as the 1D system were applied. No flow BC and hydrostatic BC are applied on the left and right boundary, respectively. The model domain is initially saturated with water while SC CO_2 is injected at the left-most node at a depth of 165m with a constant rate of 500Mt/yr for 10 years. In the 2D simulations, the seal and reservoir are assigned with isotropic permeabilities of 0.1 and 100 mD, respectively. A uniform porosity of 0.15 is assigned throughout the model domain. The locations of the 4 seal regions are list in Table 3, other portions of the domain are the reservoir formation. There are a total of 8 layers of seal rock and 120 layers of formation rock.

Table 3 Location of shell portion in 2D simulations

Shell regions Z direction	Grid region (x direction all from 1 to 100)				
	Region 1	Region 2	Region 3	Region 4	other parts are formations
	21-22	43-44	65-66	87-88	

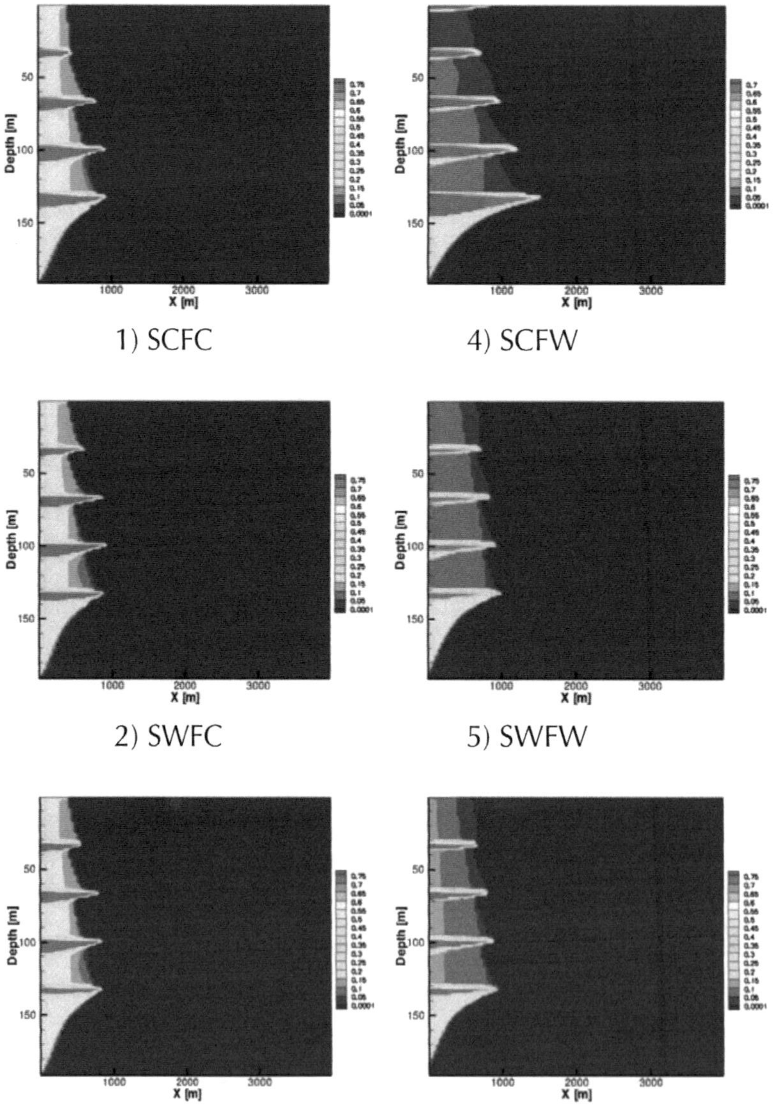

1) SCFC 4) SCFW

2) SWFC 5) SWFW

3) SXFC 6) SXFW

Figure 4. SC phase saturation at 100 years in 2D simulations. The injections start at 0.0 year and last for 10 years.

Six simulations were conducted to investigate the influence of wettability on the migration and distribution of the SC phase plume. The wettability of seal and formation rocks are listed in Table 4, in which the symbol S and F specify seal and formation rocks, the symbol W and C indicate that the rock is water or SC CO_2 wetted. The symbol X indicates that the contact angle is a function of dissolved CO_2 concentration, according to Equation (6).

Table 4 Location of shell portion in 2D simulations

Case	1	2	3	4	5	6
wettability	SC,FC	SW,FC	SX,FC	SC,FW	SW,FW	SX,FW

Figure 4 shows saturation profiles at 100 years for the 6 scenarios investigated. The first 3 cases with SC CO_2 wetted formation rocks (Category FC) are shown in the left panel in Figures 4; the 3 cases with water-wetted formation rocks (Category FW) are shown on the right panel. Generally, the SC phase saturations in the formations shown in the left panel are higher than those shown in the right panel, which is consistent with examples presented in the previous section. Also within each panel, the SC phase saturations in the seals with SC CO_2 wetted rocks are higher with CO_2 spreading further horizontally than those with water-wetted rocks.

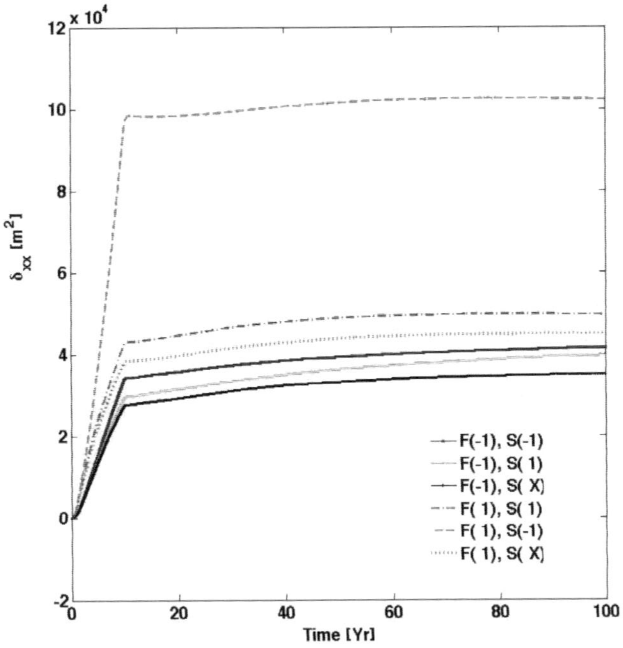

Figure 5. x-diversion of SC plume in 2D simulations.

The horizontal spreading distance can be characterized by the diversion in the X direction δ_{XX} of the SC phase plume, defined as $\delta_{XX} = M_{20}/M_{00}$. Figure 5 shows δ_{XX} for the 6 cases studied. The 3 cases of Category FW have greater δ_{XX} values than the 3 cases of Category FC. The highest δ_{XX} is more than 3 times greater than that of the smallest value, indicating the strong influence of wettability. Within each category, δ_{xx} decreases in the order of SC>SW>SX. The δ_{xx} for SW is greater than that for SX, a result of capillary pressure, similar to the behavior shown in Figure 3b.

Concluding remarks

In this study, we investigated the effects of wettability on the migration and distribution of a SC CO_2 plume during CO_2 geological storage processes. It was found that SC CO_2 has a higher saturation in SC CO_2 wetted regions than that in water-wetted regions, which indicates higher efficiency of SC CO_2 wetted formations for storage purposes. Capillary pressure acts as a barrier while the SC CO_2 migrates from SC CO_2 wetted regions to water-wetted regions. The effects of wettability and contact angle variation have a large influence on the migration and spatial distribution of the SC CO_2 plume, and should be taken into consideration in future numerical simulations.

References

1. Holt, T., I.J. Jensen, and E. Lindeburg, *Underground storage of CO2 in aquifers and oil reservoirs.* Energy Conversion & Management, 1995. **36**: p. 535--538.
2. Karbol, R. and A. Kaddour, *Sleipper Vest CO2 disposal -- Injection of removed CO2 into the Utsira formation.* Energy Conversion & Management, 1995. **36**: p. 509--512.
3. Bachu, S., W.D. Gunter, and E.H. Perkins, *Aquifer disposal of CO2: Hydrodynamic and mineral trapping.* Energy Conversion & Management, 1994. **35**(4): p. 269--279.
4. Hitchon, B. and (Eds.), eds. *Aquifer Disposal of Carbon Dioxide*. 1996, Geoscience publishing: Sherwood Park, Alberta, Canada.
5. Orr Jr., F.M., *Storage of carbon dioxide in geologic formations.* Journal of Petroleum Technology, 2004. **56**: p. 90--97.
6. Bachu, S. and B. Bennion, *Effects of in-situ conditions on relative permeability characteristics of CO2 brine systems.* Environ. Geol., 2007.
7. Chalbaud, C., et al., *Interfacial tension measurements and wettability evaluation for geological CO2 storage.* Advances in Water Resources, 2009. **32**: p. 12.
8. Bradford, S.A., L.M. Abriola, and K.M. Rathfelder, *Flow and entrapment of dense nonaqueous phase liquids in physically and chemically heterogeneous aquifer formations.* Advances in Water Resources, 1998. 22(2): p. 16.
9. Paker, J.C., Lenhard, R.J., and Kuppusamy, T. A parametric model for constitutive properties governing multiphase flow in porous media, Water Resource Research 1987.23(4): p. 618.
10. Lu, C. and P.C. Lichtner, *High resolution numerical investigation on the effect of convective instability on long term CO2 storage insaline aquifers.* J. Phys.: Conf. Ser., 2007. 78.
11. Christ, J.A., L.D. Lemke, and L.M. Abriola, *Comparison of two-dimensional and three-dimensional simulations of dense nonaqueous phase liquids (DNAPLs): Migration and entrapment in a nonuniform permeability field.* Water Resource Research, 2005. **41**: p. W01007.

Meeting the challenge of making use of advanced chemical engineering models in oil and gas production

Michel Pons
technologyofficer@colan.org
CAPE-OPEN Laboratories Network
c/o IFP, 1&4 avenue de Bois Préau, 92500 Rueil-Malmaison, F-92500, France

Summary

The accurate design of oil and gas production systems as well as the safe and optimized operation of such systems rely more and more on modelling techniques that bring together physico-chemistry, multiphase flow and thermodynamic advances. The progress made takes the form of new closure laws, i.e. new hydrodynamic point models, of new thermodynamic models representing better for example hydrate or wax formation and also of new unit operation models that better describe certain pieces of equipement.

For these advances to be experimented and then routinely used, it is necessary to make them readily available in the software tools used for the design and operation of production systems. The availability has to be quick and widespread. Interoperability is then the key to achieve this.

CAPE-OPEN is a freely available interoperability standards for process simulation tools. It has been implemented in many major commercial process simulation tools. A number of key providers of advanced modelling technologies for multiphase flow and thermodynamics targeted at oil and gas production have already moved to CAPE-OPEN in order to expand the accessibility to these technologies. It is a strategic option to be considered in the early stages of any new model development.

Context

Research and development activities that support oil and gas production systems result often in pieces of software being developed. Take as example the output from the various programs at Tulsa University in the US, or the output of SINTEF in Norway, or the output of IFP Energies Nouvelles in France. Operators have the need to make use of these results in the various tools that support design and operation.

Too often software tools are made up as stand-alone pieces of software with their proprietary format for inputs and outputs, including also often the need to enter thermodynamic data instead of relying on 3rd party thermodynamic servers. Such limitations lower the usability of a number of R&D results and hinder their application in operation.

CAPE-OPEN standards have been developed by industry (operators as well as software vendors) in order to mitigate the above mentioned limitations by proposing a seamless interoperability between software components used in process simulation.

CAPE-OPEN

The development of a plug and play standard in process simulation software results from the initiative of a group of operating companies in the mid 90s. The development took the form of a series of European Union (EU) funded projects, led by IFP, starting with CAPE-OPEN (Jan 1, 1997 till June 30, 1999) and followed by Global CAPE-OPEN (July 1, 1999 till March 31, 2002) and GCO Support projects. These projects gathered operating companies, software vendors and academics. Global CAPE-OPEN was also labeled as an Intelligent Manufacturing Systems project facilitating the involvement of industrial and academic partners from Canada, US and Japan.

The CAPE-OPEN and Global CAPE-OPEN projects developed a set of interface standards that enable interoperability between the various pieces of software making up process simulation tools. The GCO Support project delivered a set of disseminating means for CAPE-OPEN standards.

Following the end of the EU projects, the CAPE-OPEN (CO) standards are now maintained and developed by the CAPE-OPEN Laboratories Network (CO-LaN), founded in 2001 by a group of twenty end user organizations, universities and software vendors. CO-LaN is funded by process industries end user companies (1). CO-LaN counts nowadays more than 80 members.

The CAPE-OPEN standards are covering the whole range of software components making up any process simulator. The standards are made of a documentation set describing the CAPE-OPEN interfaces for each type of component. The Thermodynamic and Physical Properties interface specification as well as the Unit Operation interface specification have been implemented in numerous Process Modelling Environments (i.e. process simulators) and Process Modelling Components (i.e. thermodynamic servers, unit operations, numerical solvers, etc…).

This paper discusses how these two sets of interfaces are of interest and of use to the oil and gas industry.

Thermodynamic servers

Modelling of oil and gas production systems relies often on black oil thermodynamic models but is also more and more relying on a full compositional approach that calls for advanced thermodynamic models.

While advanced thermodynamic models may be found in existing commercial software such as MultiFlash from Infochem Computer Services, an issue is to have them readily available when needed. While Infochem offers for MultiFlash a proprietary interface to connect it to any other piece of software, Infochem has recognized early on that making MultiFlash CAPE-OPEN compliant is widening its market applicability. Hence MultiFlash is distributed as a CAPE-OPEN compliant thermodynamic server that can be readily used in all CAPE-OPEN compliant process simulators (Aspen Hysys, PRO/II, UniSim Design, Aspen Plus, gPROMS, ProSimPlus, COFE, ChemCad, etc…).

The Center for Energy Resources Engineering (CERE) at Denmark Technical University has developed two thermodynamic models that have been encapsulated (Breil et al., 2007a, 2007b) as CAPE-OPEN compliant pieces of software:

- CPA: an extension to the SRK equation of state
- PC-SAFT

CERE recognizes as advantage of CAPE-OPEN that using such a technology its products are applicable to more than just one process simulator. Since CERE serves a consortium of many companies, not all of them using the same process simulator, it exemplifies the need for a common and immediate interfacing technology.

Recently Process Systems Enterprise has revealed its development of gSAFT, a thermodynamic component relying on the SAFT technology. This specific SAFT technology has been developed and is still under development at Imperial College. At PSE Annual Meeting 2010 on September 17, 2010, Process Systems Enterprise has announced that it is considering releasing gSAFT as a CAPE-OPEN thermodynamic component to ensure it is used not only in gPROMS but also in other process simulators, providing thermodynamic consistency within any end-user company on thermodynamics. gSAFT is currently under validation by a major oil&gas company.

Rühr Universität Bochum department of thermodynamics has developed the GERG 2004 equation of state. In order to make use of this reference model, the department of thermodynamics (Dauber et al. 2010) has decided to make available this model as a CAPE-OPEN Property Package dedicated to the modelling of fluids in a LNG plant. It has already permitted btroubleshooting studies to be conducted in a variety of process simulators.

The above shows that academic departments as well as university research consortia or software vendors can take advantage of CAPE-OPEN technology to bring to the market i.e. to the end-users results that may bring advantage to the modelling of oil and gas production systems.

Unit Operations

Oil and gas production systems rely on pieces of equipment that have been specifically designed for that domain. Also oil and gas production systems are made of pipes in which flow assurance problems have to be tackled.

IFP Energies Nouvelles has teamed up with TOTAL in order to develop a pipe unit operation model based on the TACITE project on hydrodynamic modelling. In order to make use of this pipe model in various process simulators, it was decided to make this unit operation CAPE-OPEN compliant. Subsequently this unit operation has been plugged into various process simulators such as ProSimPlus, PRO/II, Aspen Hysys and Indiss Plus to conduct for example optimization studies as in Floquet et al. (2009).

TOTAL, as member of the Tulsa University Paraffin Deposition Project, had access to the TUWAX software that serves partly as knowledge repository for the experimental and modeling work conducted on paraffin deposition. TOTAL felt the need to combine TUWAX with MultiFlash in order to benefit from recent advances in paraffin thermodynamic modelling. Also TOTAL felt the need to include the TUWAX Unit Operation in models of oil and gas production systems developed in process simulators. Making TUWAX a CAPE-OPEN compliant piece of software fulfilled these needs (Ricordeau et al., 2007).

TOTAL, as member of the Tulsa University Separation Technology Projects, had access to the Gas Liquid Cylindrical Cyclone (GLCC) design and verification code. TOTAL felt the need to use such a unit operation model within models of oil and gas production systems developed within process simulators. Making TUWAX a CAPE-OPEN compliance piece of software fullfilled this need permitting use of the GLCC Unit Operation in PRO/II and Aspen Hysys (Gassies et al. 2008).

HYSIFLO, an IFP Joint Industry Project, has the objective to develop a Hydrate Flow Module, i.e. a predictive tool that will make easier for process engineers to evaluate flow properties and the risk of blockage when transportation of the produced fluid in pipelines is in the hydrate domain. The Hydrate Flow Module is delivered as a CAPE-OPEN Unit Operation so that it may be plugged in process simulators and used together with advanced thermodynamic servers such as MultiFlash from Infochem Computer Services Ltd.

The above examples confirm that developing and deploying CAPE-OPEN Unit Operations for the oil and gas industry is working.

Process simulators

In order to make use of the above cited thermodynamic servers and unit operations, CAPE-OPEN compliant process simulators are needed. Aspen Hysys from Aspentech, PRO/II from SimSci-Esscor, UniSim Design from Honeywell, gPROMS from Process Systems Enterprise, ProSimPlus from proSim SA are among the main commercial process simulators used when modelling in steady-state oil and gas production systems.

Together with IFP, within the research project TINA (see Bratland, 2009), TOTAL backed development of the INDISS software platform into INDISS Plus (Gainville et al., 2007) with the aim to permit model re-usability, to easily add in-house and specific models, to seamlessly use 3rd party software, to easily switch from one model to another. INDISS Plus is developed and marketed by RSI Simcon.

Kongsberg Oil & Gas Technology has mentioned its willingness to enhance its K-Spice dynamic modular simulator with CAPE-OPEN interfaces. First, K-Spice will be able to host CAPE-OPEN compliant units as well as thermodynamic property packages. In addition, it will be possible to plug the complete K-Spice simulator as a unit into some other flowsheet.

The above shows that while it is possible to develop new process simulators exhibiting additional functionalities, and that CAPE-OPEN compliance is making up the future of existing process simulators, to better serve their clients.

Conclusion

CAPE-OPEN is providing the interoperability that permits the rapid and complete use of results of research and development activities for the oil and gas industry. CAPE-OPEN has already been an important feature of numerous projects and is gathering momentum. It should be considered as a desirable feature for any software development involved in the modelling of oil and gas production systems.

References

Belaud J.P. et al. (2003), Proceedings of Foundations Of Computer-Aided Process Operations, Coral Springs, Florida, USA.

Bratland O. (2009) Update on Commercially Available Flow Assurance Software Tools: What they can and cannot do? PetroMin PIPELINER, Jan-Mar 2009. presented at 4th Asian Pipeline Conference and Exhibition.

Breil M., Kontogeorgis G., von Solms N. and E. H. Stenby (2007a). Use your own models in process simulation. CAPE-OPEN - an international standard. Journal of Chemical Engineering.

Breil M., Kontogeorgis G., von Solms N. and E. H. Stenby (2007b). CAPE-OPEN: An International Standard for Process Simulation. Chemical Engineering (ISSN: 0009-2460) , vol: 114, issue: 13, pages: 52-55, 2007, Access Intelligence, LLC

Dauber F., Span R. and P. Schley (2010) Influence of thermodynamic property models on LNG process simulation. International issue 2010, gwf-Gas/Erdgas, pp. 2-4.

Floquet P., Joulia X., Vacher A., Gainville M. and Pons M. (2009) Numerical and Computational Strategy for Pressure-Driven Steady-State Simulation of Oilfield Production. Computers & Chemical Engineering, vol. 33 (n° 3). pp. 660-669. ISSN 0098-1354.

Gainville M., Roux P., M. Pons, Ricordeau A. and D. Paen (2007). Integrated Flow Modeling Platform Using CAPE-OPEN standard. Paper 110864-MS, SPE Annual Technical Conference and Exhibition, 11-14 November 2007, Anaheim, CA.

Gassies M., Ricordeau A, Mohan R.S.and O. Shoham (2008). GLCC CAPE-OPEN compliant Unit Operation. Paper 397, *AIChE Annual Meeting 2008*, Philadelphia, PA.

Ricordeau A., Sarica C. and A. Mathieu (2007). Development of TUWAX as a CAPE-OPEN compliant process modelling environment. Paper 459c, *AIChE Annual Meeting 2007*, Salt Lake City, UT.

(457b) Comparison of Gas Production From Gas Hydrate Reservoirs Using Vertical and Horizontal Wells

Manohar Gaddipati, Lavanya Nyayapathi and Brian J. Anderson, Department of Chemical Engineering, West Virginia University, Morgantown, WV

Methane hydrates may contain significant amounts of hydrocarbons both in the permafrost and in oceans due to favorable settings of pressure and temperature. This simulation-based work is aimed to estimate and compare natural gas production from hydrate-bearing reservoirs on the Alaskan North Slope using horizontal and vertical wells in hydrate bearing reservoirs. It has been established that high amounts of gas can be obtained from hydrate reservoirs using depressurization technique and vertical wells [1]. In this paper, numerical studies are conducted on similar reservoirs using horizontal and vertical wells in the hydrate bearing layers using CMG STARS. CMG STARS (Steam, Thermal, and Advanced Processes Reservoir Simulator) is a three-phase multi-component thermal and steam additive simulator, developed by the Computer Modeling Group Limited (CMG). This paper also examines the feasibility of hydraulic fracturing in the permafrost methane hydrate reservoirs.

The geologic settings of the hydrate reservoirs are obtained from the USGS-NETL Methane Hydrate Code Comparison project [2]. For comparison of performance of both the wells, various values of porosity, intrinsic permeabilities, reservoir temperature, hydrate saturation and free water saturation will be used in the simulations and results will be compared. In the base reservoir model, a hydrate layer is bound by two shale layers, above and below. The vertical well simulation will be carried out using a cylindrical domain and the horizontal well simulation will be carried out using a rectangular Cartesian domain. Average pressure and temperature are 12°C and 9 MPa in the hydrate bearing layer. The production will be adjusted to the volume of hydrate present in each of the reservoirs. It is predicted that the horizontal well will produce gas more instantaneously than the vertical well.

1. Sloan, E.D., and C. Koh, *Clathrate Hydrates of Nautral Gases*. 3rd Edition, Boca Raton, FL: Taylor and Francis, Inc., 2008

2. Wilder, J., et al. *An International Effort to Compare Gas Hydrate Reservoir Simulators*. In *Proceedings of the 6th International Conferenceon Gas Hydrates*. 2008. Vancouver, British Columbia, Canada.

See more of this Session: Chemistry, Thermodynamics and Multiphase Flow of Complex Fluids Related to Flow Assurance I
See more of this Group/Topical: Topical D: Chemical Engineering in Oil and Gas Production and Other

(457c) Morphological Investigation of Cyclopentane Hydrates: Effect of Cyclopentane Activity and Surface Active Agents

Prasad U. Karanjkar[1], Jae W. Lee[2] and Jeffrey F. Morris[1], (1)Benjamin Levich Institute and Department of Chemical Engineering, The City College of New York, New York, NY, (2)Department of Chemical Engineering, The City College of New York, New York, NY

Clathrate hydrates, which are generally termed gas hydrates when natural gas is a guest molecule, generate interest because of the potential negative consequences of hydrate plug formation for transport of natural gas and hydrogen, and because of their potential application in CO2 sequestration, or as phase change materials in refrigeration. Proper analysis of hydrates in any scenario will require knowledge of the dynamic processes of formation and dissociation as well as understanding of their rheological and morphological properties. This work addresses the last one of these issues related to model oil-field emulsions. Cyclopentane forms clathrate hydrates at atmospheric pressure and is used as the guest molecule in hydrates. Morphological behavior of a single water drop immersed in a hydrocarbon phase containing cyclopentane under hydrate forming conditions is presented. It has been found that the concentration of cyclopentane in the external hydrocarbon phase affects the thermodynamics and kinetics of the process as well as micro-structural characteristics of hydrates. The equilibrium dissociation temperature and the rate of hydrate formation decreases with a drop in cyclopentane activity (concentration). Hydrate formation preferentially occurs at the water-hydrocarbon interface leading to formation of a hydrate shell around a water drop. Generally crude oils carry a significant amount of natural surfactants, e.g. asphaltenes, inspiring us to study the effect of surface active agents over the hydrate morphology. Results are reported for different surfactants present only in either the hydrocarbon or water, or in both the phases; naturally-occurring surface active agents from crude oil have also been studied. Our results indicate that head group area of a surfactant significantly affects the sequence of events in hydrate formation, and thus modifies the final morphology. Bulky head group surfactants present in the hydrocarbon phase lead to a mushy / dendritic morphology while a faceted / polycrystalline shell is found in their absence. Interaction of two water drops during the process of hydrate formation possesses a significant value to understand a global emulsion behavior during plug formation and results on this problem will be described.

See more of this Session: Chemistry, Thermodynamics and Multiphase Flow of Complex Fluids Related to Flow Assurance I

See more of this Group/Topical: Topical D: Chemical Engineering in Oil and Gas Production and Other Complex Subsurface Processes

(457d) Prediction of Asphaltene Deposition in Well Bores

Anjushri S. Kurup, Department of Chemical and Biomolecular Engineering, Rice University, Houston, TX, Walter G Chapman, Chemical & Biomolecular Engineering, Rice University, Houston, TX, Francisco M. Vargas, Chemical Engineering, The Petroluem Institute, Abu Dhabi, United Arab Emirates, Jefferson Creek, Chevron Petroleum Technology Corporation, Houston, TX, Jianxin Wang, Chevron, Houston, TX, Jill Buckley, New Mexico Institute of Mining and Technology, Socorro, NM and Hariprasad J. Subramani, Flow Assurance, Chevron Energy Technology Company, Houston, TX

In the energy industry, asphaltenes are considered as potential damage creators. As the physical or chemical conditions inside the reservoir or the well bore change, asphaltenes that are originally stable in the oil phase, tend to precipitate, aggregate and / or deposit on the containing structures. This results in clogged pipes, well bores and other production equipments such as pumps. Due to this notorious behavior, asphaltenes are often termed as "the cholesterol of petroleum". The potential damage produced by asphaltene deposition during oil production has motivated the development of several experimental techniques and theoretical models, trying to understand and predict the asphaltene behavior under various conditions. Successful application of the Perturbed Chain version of the Statistical Associating Fluid Theory Equation of State (PC-SAFT EOS) in modeling the asphaltene phase behavior under both ambient and reservoir conditions have already been reported. The effect of factors such as pressure, temperature, composition and asphaltene poly dispersity and the effect of field conditions such as injection of CO2, oil based mud contamination and commingling of oils on the phase behavior of asphaltenes, have been described in the literature. Although important progress has been made in this area, the ability to predict deposition and possible pipe or well bore blockage due to asphaltenes, is still lacking. In this work, we present the development of a simulation tool that can potentially predict the amount of asphaltene deposition in well bores. The transport equations are coupled with kinetic models to describe the precipitation, aggregation and deposition of asphaltenes. The phase behavior of asphaltenes in oil is described by using the PC-SAFT EOS. The model predictions are compared with experimentally obtained asphaltene deposition profiles inside a capillary tube. Sensitivity analyses to understand the effect of various operating transport and kinetic parameters on the deposition profile is also presented. The effect of temperature variation inside the well bore, as the oil flows up, is also studied. Certain key aspects that could possibly improve the model predictions are also discussed. The development of a deposition simulator such as described above will certainly aid in a better understanding and prediction of asphaltene precipitation and deposition in well bores in presence of varying operating conditions. Identifying the conditions that can potentially create asphaltene problems will help operators not only to have better production planning, so as to avoid costly mistakes, but can also aid in improved new well bore designs.

See more of this Session: Chemistry, Thermodynamics and Multiphase Flow of Complex Fluids Related to Flow Assurance I

(457e) Investigations Into Asphaltene Stability Using Capillary Deposition

Michael P. Hoepfner, Tabish Maqbool and H. Scott Fogler, Chemical Engineering, The University of Michigan, Ann Arbor, MI

Asphaltene deposition is a costly concern in flow assurance. Precipitation and deposition can occur in the porous rock formations, well-bore, production lines and in the refinery. Recent work in asphaltene aggregation at low precipitant concentrations has revealed that asphaltenes are less stable than previously thought [1]. These measurements employed optical microscopy to detect asphaltene instability in batch experiments. Detection of instability can take months with this setup when the precipitant concentration is low.

A capillary deposition apparatus has been developed that can test for asphaltene instability at low precipitant concentrations in short periods of time. This configuration is useful because it does not require asphaltene aggregates to grow to a size of ~0.5 microns before detection is possible. There is a continuous supply of unstable asphaltenes that are able to form a deposit on the capillary wall. The asphaltene deposit causes an increase in the differential pressure across the capillary test section. In addition to the detection of instability, capillary deposition can provide insight into the deposition/ stability mechanism for asphaltenes. The aggregation and deposition of asphaltenes at low precipitant concentrations questions the applicability of previous experiments and thermodynamic models to predict asphaltene stability.

[1] T. Maqbool, A.T. Balgoa, and H.S. Fogler, "Revisiting Asphaltene Precipitation from Crude Oils: A Case of Neglected Kinetic Effects," Energy & Fuels, vol. 23, Jul. 2009, pp. 3681-3686.

See more of this Session: Chemistry, Thermodynamics and Multiphase Flow of Complex Fluids Related to Flow Assurance I
See more of this Group/Topical: Topical D: Chemical Engineering in Oil and Gas Production and Other Complex Subsurface Processes

(457f) Coalescence of Asphaltene Stabilized Water Droplets

David Harbottle, Jayant Rane and Sanjoy Banerjee, The Energy Institute, The City College of The City University of New York, New York, NY

A key processing step in the recovery of crude oil is the break-up of the stable water-in-oil (w/o) emulsion in large settling tanks. Typically these continuous flow settlers operate with a residence time between 10 and 15 minutes, reducing the water cut to below 5%. To aid separation and maintain process throughput, droplet growth is promoted in the pipeline leading up to the settler. It is generally believed that droplets smaller than 100μm at the settler inlet are carried over in the oil phase. If a significant quantity of the inlet water is carried over, the efficiency of the secondary stage cleaning process (commonly electrocoalescence) is reduced; hence the growth of sub 100μm droplets prior to separation and their removal in the settler is crucial for maintaining throughput.

To increase the droplet size from tens to hundreds of microns, chemical additives are pumped into the pipeline prior to the gravity settler. These additives (demulsifiers) act to penetrate a stable water-oil interface and allow for the coalescence of water droplets when in intimate contact. Without the addition of demulsifiers, surface active components such as, asphaltenes and resins form a stabilizing film around the water droplet preventing coalescence.

Demulsifier action and their effect under different flow conditions is still not a well understood science. In the current study 3 demulsifiers (IGEPAL CO – 520, 720 and 890) have been chosen to investigate the effect of (i) molecular weight and (ii) concentration on droplet coalescence. In addition, the fluid shear rate and the time of shear are also investigated.

'Model' emulsions have been prepared with a 30% water cut dispersed in a light mineral oil. The water droplets are prepared in the size range of 10-30μm and stabilised by an asphaltene film. The asphaltenes are extracted from a crude sample recovered from the Grane oil field. A flow wheel (volume 2 liters) is used to shear the droplets, with 30ml samples removed and drained through the Dispersion Technology (DT-1200) acoustic particle size analyzer to measure the droplet size distribution. Droplet size distributions are compared before and after shearing to investigate the effect of both chemical addition and fluid shear on the coalescence rate.

Preliminary studies in the absence of demulsifiers have shown that for stable emulsions there is a critical flow range where droplet coalescence is observed. This study highlights the importance of chemical dosing and fluid shear when breaking stable w/o emulsions.

See more of this Session: Chemistry, Thermodynamics and Multiphase Flow of Complex Fluids Related

USE OF ANHYDRIDE BASED POLYMERS AS SCALE CONTROL ADDITIVES IN PRODUCTION OF HEAVY OILS
AIChE 2010 FALL PROCEEDINGS - Paper # 524a

Y. A. Roomi, K. F. Hussein and M. R. Riazi, Kuwait University, Kuwait*
**Corresponding author: RiaziM.com, email: mrriazi@gmail.com*

Introduction

Waterflooding is one of the procedures used in petroleum production. Recent studies show that even for high viscosity heavy oil fields waterflooding is the most inexpensive method of oil recovery in comparison with more expensive thermal EOR methods [1]. Many countries now contain significant amounts of heavy oils in their reserves [2,3]. During such processes inorganic chemicals in water may precipitate forming scales which may cause major problem in flow assurance and oil productivity. Precipitation of inorganics in water such as calcium sulfate ($CaSO_4$) or calcium carbonate ($CaCO_3$) is referred as scaling. In oil production by waterflooding as water approaches the vicinity of production well, initial pressures are diminished in which calcium sulfate solubility is rapidly reduced and scales are formed in the vicinity of wellbores causing restrictions in flow of oil from production well [4,5]. Inorganic scales are also a major problem in flow assurance in offshore oil and gas production which may lead to significant reductions in hydrocarbon production and costly workovers [6]. Researchers in recent years have studied mechanism of scale formation in hydrocarbon production wells [7-9].

The main method of preventing scale formation in oil and gas production or water desalination facilities is use of chemical inhibitors known as antiscalants or scale control additives (SCA). These additives are usually polymer type materials which are added in different dose rates to brine to combat scaling problems. The efficiencies of these antiscalant additives depend on the brine chemical constituents, chemical nature of antiscalant, design configuration, temperature and its reactivity with the brine. The exact chemical composition of a SCA is usually guarded secret of the manufacturer. The role or mechanism of antiscalants is still not clear in spite of their popular use [10]. Synthesis and properties of some of these polymer type SCAs have been investigated by both researchers and chemical manufacturers [11-17].

In this work we continue our previous research works on synthesis of some maleic anhydride based polymers as efficient antiscalants. Polymers based on maleic anhydride are particularly effective in sea water desalination and scale removal in oil production wells. The presence of both an electron-acceptor double bond and an anhydride moiety confers unique properties to maleic anhydride, providing an opportunity for application in chain and step-growth polymerization systems. Maleic anhydride is a widely available local material which is cost effective in comparison with other materials [18,19].

Synthesis and Experiments

Reaction schemes to produce monopolymer and copolymers from maleic anhydride (MA) in our laboratory are demonstrated in Figs. 1 and 2, respectively. Production of copolymer occurs at two step in which at first step n-octyl-acrylamide (AAO) monomer is produced as shown in Fig. 2.

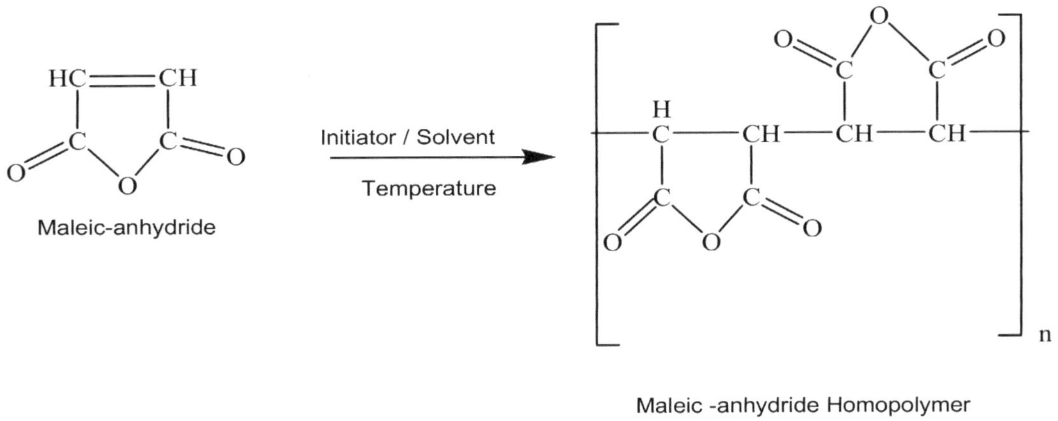

Figure 1. Homopolymer production from MA

Series of Low molecular weight maleic –anhydride homopolymers and copolymers synthesized in our laboratory are depicted in Tables 1 and 2. Maleic anhydride (Sigma, 99%) was recrystallized from benzene and dried under vacuum at room temperature. N- octyl amine (Aldrich, 98%) and acrolyl chloride (Aldrich, 96%) were used without further purification. BPO (Aldrich, 97 %) and AIBN (Aldrich, 97%) were crystallized from methanol prior to use. Molecular weight measurements of the monomers and copolymers were carried out by high temperature size exclusion chromatography (SEC). As shown in Table 1, when initiator composition was at 10 mol%, homopolymer with maximum molecular weight was produced. Table 2 shows the conditions at which maximum conversion can be achieved (50% MA, 50% AAO and 7.5% initiator).

Table 1: Conditions and Results for the Homo-polymerization of MA

Entry	Scale Inhibitor	MA (moles/liters)	Solvent	Initiator	Initiator composition, (mole %)	Conversion %	M_w[c]
1	YMR 41	0.247	O-Xylene	AIBN[a]	5	5.41	265
2	YMR 44	0.247	O-Xylene	AIBN	7.5	16.91	400
3	YMR 50	0.247	O-Xylene	AIBN	10	80.68	810
4	YMR 39	0.247	O-Xylene	AIBN	12	72.06	773
5	YMR 34	0.247	O-Xylene	AIBN	15	71.22	457
6	YMR 40	0.247	O-Xylene	BPO[b]	5	21.52	641
7	YMR 53	0.247	O-Xylene	BPO	7.5	60.64	706
8	YMR 47	0.247	O-Xylene	BPO	10	70.01	847
9	YMR 35	0.247	O-Xylene	BPO	12	30.31	405
10	YMR 31	0.247	Toluene	BPO	10	44.59	430

[a] AIBN = Azoisobutyronitrile; [b] BPO = Benzoyl peroxide; [c] Molecular weight obtained from SEC

Monomer Synthesis:

A 200 ml round bottom flask equipped with a glass stir bar was placed in an ice bath. N-Octyl amine and barium hydroxide (1%) were placed in the flask and stirred until the temperature equilibrated. Acrolyl chloride was added drop wise to the flask containing the n-octyl amine while continuously stirring. The mixture in the reaction vessel thickened after about 7 minutes indicating the formation of the monomer. The reaction mixture was then heated gradually up to 85°C, until HCl was eliminated. The resulting solution was whitish-yellow and turbid. Doubly deionized water was added to the vessel with vigorous stirring to separate the monomer. The procedure was repeated several times to eliminate excess reactants. The resulting material was dried by placing it in a 40°C oven. The waxy white colored final product was then stored at 0°C.

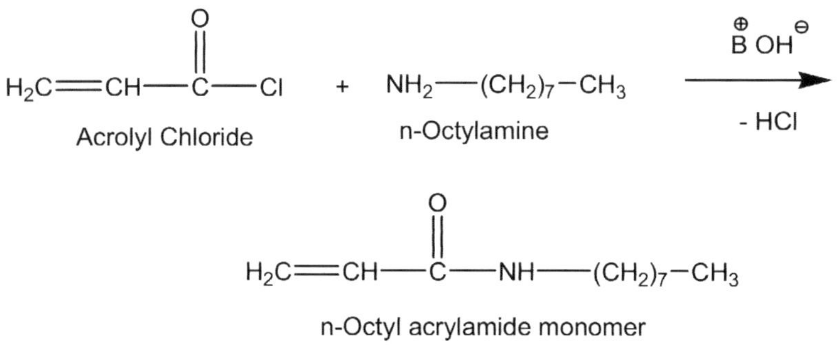

STEP- 1 : Reaction Scheme for preparation of n-Octyl acrylamide monomer (AAO)

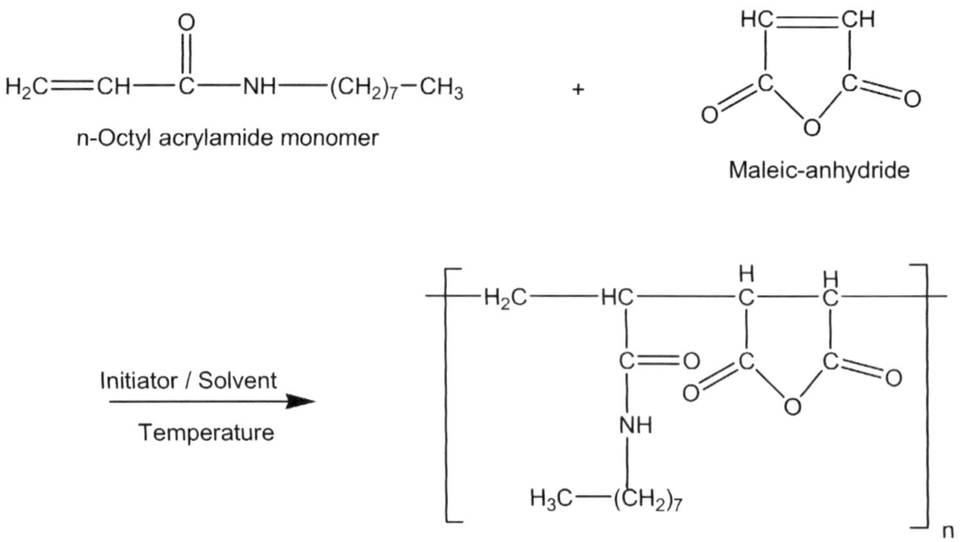

STEP- 2 : Reaction Scheme for preparation of Maleic-anhydride- n-Octyl acrylamide copolymer

Figure 2. Copolymer production in two steps

Copolymerization:

Copolymerization of MA with AAO was conducted with the objective of characterizing the reaction and to generate a series of low molecular weight co-polymers of varying composition. The reaction was conducted at four different concentrations. The samples were labeled as: MA / 30 AAO, MA / 50 AAO, MA / 60 AAO and MA / 40 AAO indicating 70:30, 50:50, 40:60 and 60:40 mol % of MA:AAO, respectively. A dry N_2 inert reaction atmosphere was maintained throughout all experiments. Two different initiators azo-isobutyronitrile (AIBN) and benzoyl peroxide (BPO), were used in the copolymerization study. The concentration of initiators (5, 7.5, 10 and 12 mol % of monomers) was varied to select and optimize their performance. The reaction temperature was maintained at the reflux temperature of o-xylene. The progress of the reaction was followed by measuring the disappearance of the monomer, which was conducted by taking out a small sample of reaction mixture every hour and determining its monomer content by LC and increase in molecular weight of the polymer by gel permeation chromatography.

A typical polymerization was carried out as follows. MA dissolved in 200ml of o-xylene was stirred and heated to reflux under a N_2 atmosphere until the temperature reached 70^0C. AAO and initiator dissolved in 50 ml of solvent were added over a period of five minutes. Refluxing was continued for a further 9 hours and the polymer was allowed to separate from the solution. Excess of solvent was distilled off and the oily layer so obtained was recrystallized in benzene to give the copolymer product. The series of n-octylacrylamide-maleic anhydride copolymers prepared by the free radical mechanism were analyzed by ^{1}H-NMR and ^{13}C-NMR spectroscopy to determine the copolymer composition [18].

Table 2: Conditions and Results for the Radical Copolymerization of MA and AAO in O-Xylene

S#	Scale Inhibitor	Composition of monomer mixture, mole%		Initiator	Initiator composition (mole %)	Conversion %	M_w/M_n	M_w[c]
		MA	AAO					
1	YMR-80	70	30	BPO[a]	7.5	49	3.89	1152
2	YMR-82	60	40	BPO	7.5	43.5	3.47	1345
3	YMR-68	50	50	BPO	7.5	52.5	1.89	1759
4	YMR-78	40	60	BPO	7.5	37.56	3.87	935
5	YMR-65	70	30	AIBN[b]	7.5	43.5	1.36	1658
6	YMR-72	60	40	AIBN	7.5	53.3	1.66	1646
7	YMR-74	50	50	AIBN	7.5	53	1.61	1890
8	YMR-75	40	60	AIBN	7.5	17	3.62	1469
9	YMR-67	50	50	BPO	5	40.2	1.63	1545
10	YMR-76	50	50	BPO	10	62	3.4	1364
11	YMR-81	50	50	BPO	12	50	3.1	1135
12	YMR-79	50	50	AIBN	5	33.25	3.78	1562
13	YMR-63	50	50	AIBN	10	53	1.67	1725
14	YMR-83	50	50	AIBN	12	43.5	3.86	1267

[a] BPO = Benzoyl peroxide; [b] AIBN = Azoisobutyronitrile; [c] Molecular weight obtained from SEC

Inhibition Efficiency of Synthesized Polymers

Synthesized polymers are all highly effective inhibitors of $CaCO_3$ and $CaSO_4$ precipitation. Precipitation tests for $CaCO_3$ were made with 0.1, 0.3 and 0.5 ppm while for $CaSO_4$ tests were conducted at 0.25, 0.5, 1, 2, 3, 5 and 7 ppm polymer concentrations. Virtually complete inhibition is achieved with 0.5 ppm polymer as it can be observed from Fig. 4 and Fig. 5 for $CaCO_3$. For homopolymers no significant difference in activity among the polymers is noted during this test as shown in Fig. 6, where as for copolymer series the inhibition percent increases with increase in molecular weight as can be observed from Fig. 7. Test methods used for inhibition efficiency of polymers were according to ASTM D 1126-02 and NACE TM 0374-95 [20,21]. To calculate the percentage inhibition of synthesized polymers following equation was used:

$$\% \text{ inhibition} = \frac{[Ca^{2+}]_{antiscalant} - [Ca^{2+}]_{blank}}{[Ca^{2+}]_{original} - [Ca^{2+}]_{blank}} \times 100$$

where:

$[Ca^{2+}]_{antiscalant}$ = concentration of soluble Ca^{2+} in an antiscalant containing solution (mg/L)
$[Ca^{2+}]_{original}$ = concentration of soluble Ca^{2+} in original solution (mg/L)
$[Ca^{2+}]_{blank}$ = concentration of soluble Ca^{2+} in a blank solution (mg/L)

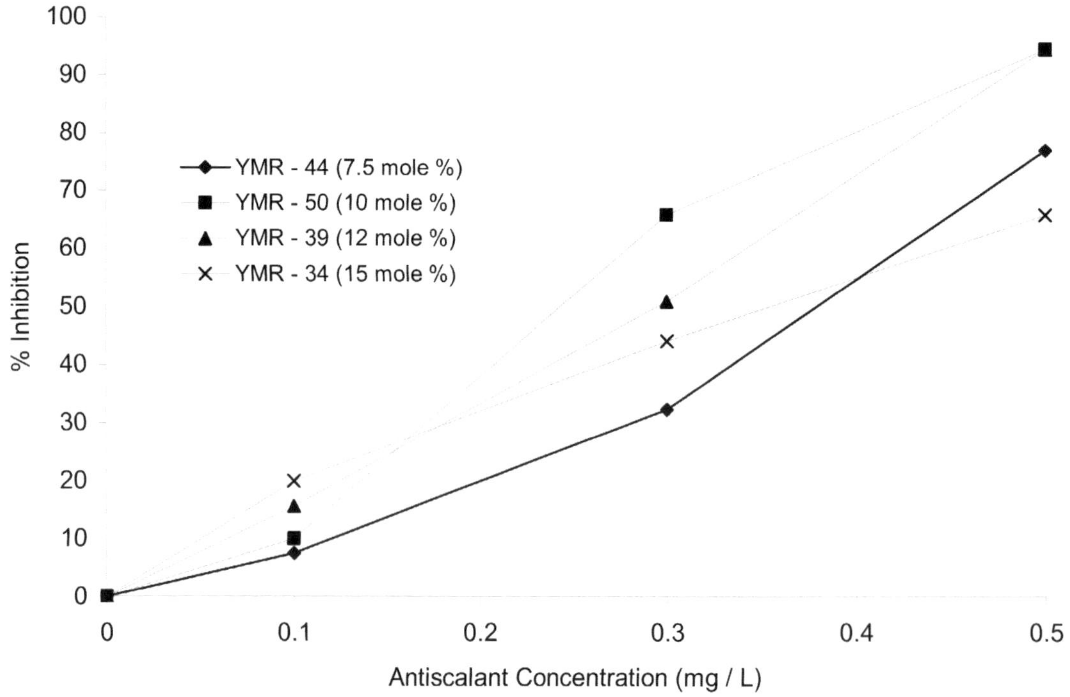

Figure 4: Inhibition% for $CaCO_3$ scales using synthesized homo-polymers obtained at varying concentration of AIBN initiator.

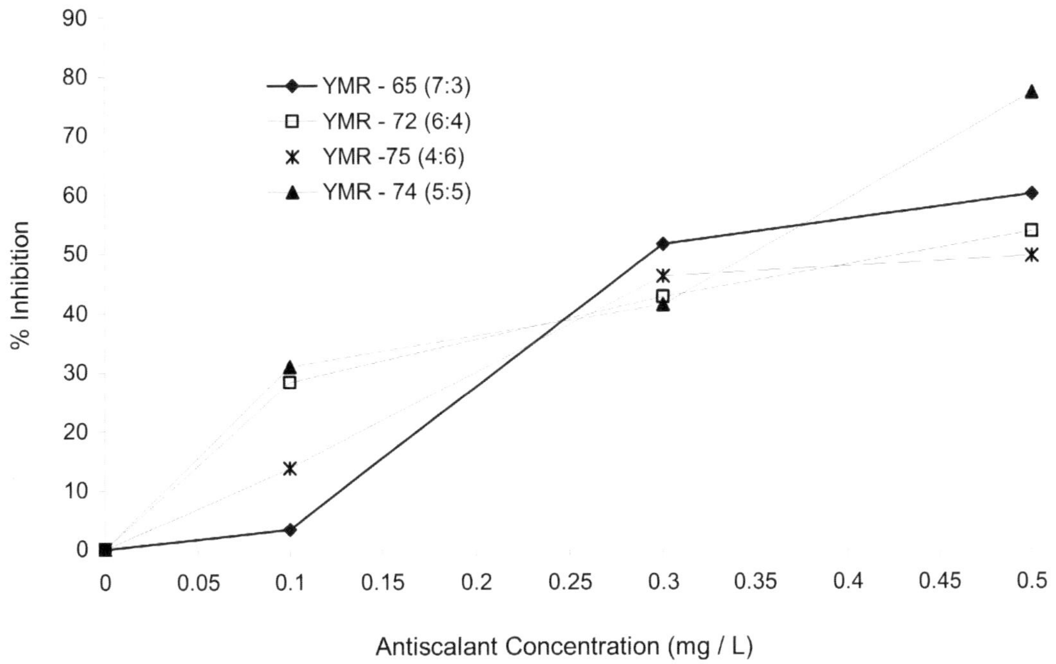

Figure 5: $CaCO_3$ Inhibition% for $CaCO_3$ scales using synthesized homo-polymers obtained at 7.5 mole % AIBN concentration and varying monomer (MA / AAO) ratio

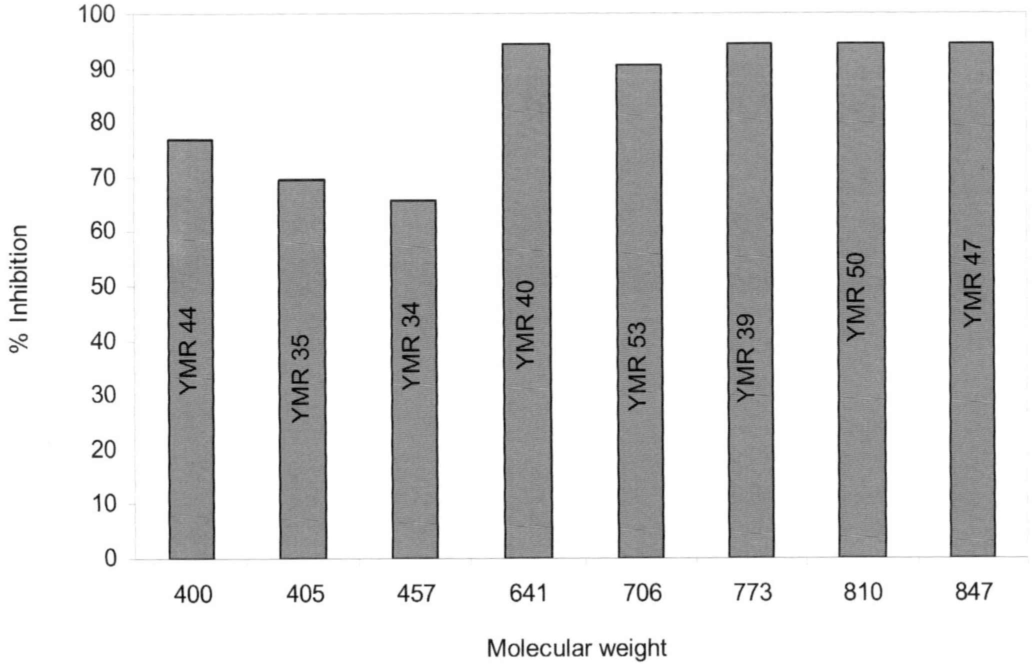

Figure 6: Inhibition % for $CaCO_3$ scales with molecular weight of MA homopolymers.

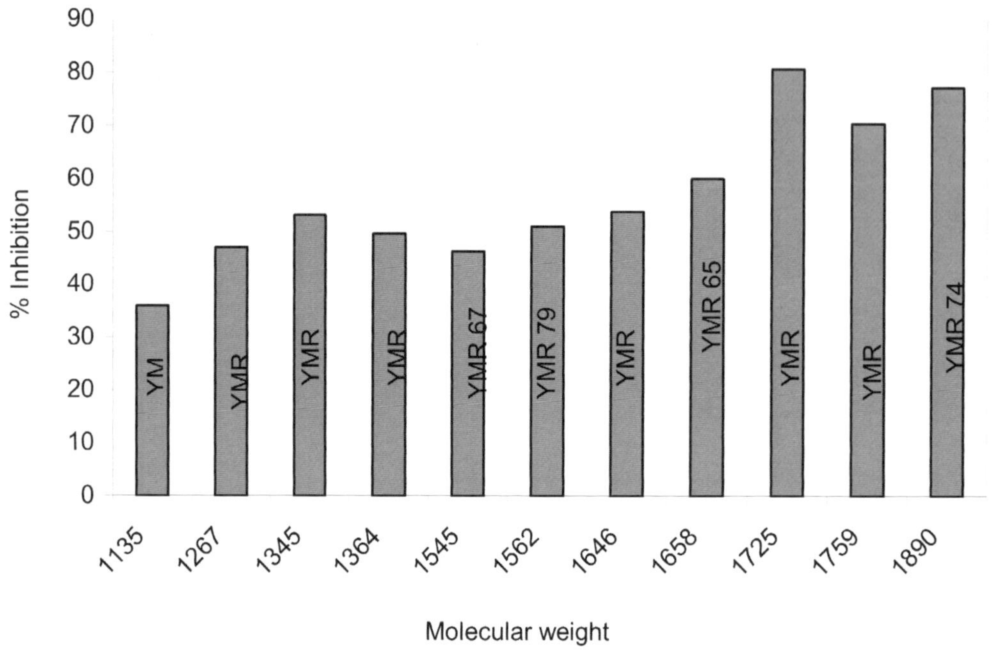

Figure 7: Plot showing increase in inhibition % for CaCO3 scales with molecular weight of the antiscalants (MA / AAO Copolymer)

To view the impact of produced antiscalants photographs of $CaCO_3$ scales before and after use of synthesized polymer are shown in Figs 8 and 9, respectively. Similar performances are observed for $CaSO_4$ and $Mg(OH)_2$ scaling.

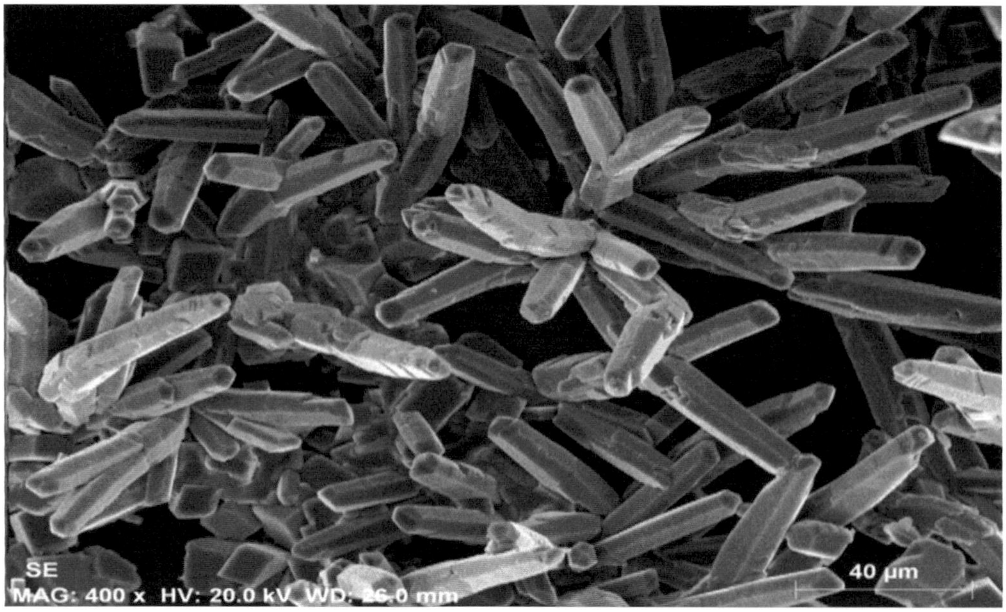

Figure 8: SEM photograph for the $CaCO_3$ in the absence of scale inhibitor (BLANK)

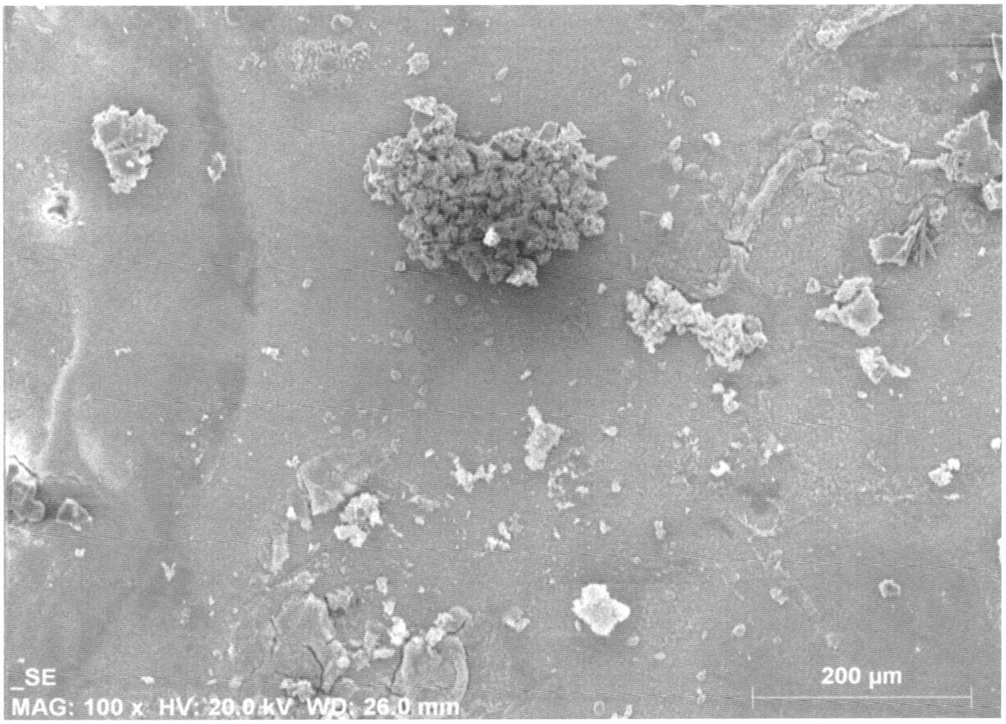

Figure 9: SEM photograph for the $CaCO_3$ in the presence of YMR-63

Conclusions

In this paper synthesis and production of series of antiscalants homo and copolymers based on maleic anhydride are presented. Produced polymers have been used as inhibitors for precipitation of calcium carbonate, calcium sulfate and magnesium hydroxide. Results of extensive experiments and testing show that synthesized polymers demonstrate very good efficiency and could be used as cost effective antiscalants in oil and gas production wells as well as in water desalination units.

Acknowledgment

The funding for this project was provided by the Research Administration of Kuwait University under Project No. RE 05/06 which is appreciated.

References

1. Mai, A., Kantzas, A. (2009), 'Heavy Oil Waterflooding: Effects of Flow Rate and Oil Viscosity,' *Journal of Canadian Petroleum Technology*, Vol 48, No. 3, pp. 42-51.
2. Riazi, M. R., (2010), 'Energy, Economy, Environment and Sustainable Development in the Middle East and North Africa', *International Journal of Oil, Gas and Coal Technology* (IJOGCT), Vol 3, No 3, pp. 201-240.

3. Riazi, M. R., (2005) "*Characterization and Properties of Petroleum Fractions*," ASTM International, Conshohocken, PA, USA. (http://www.astm.org/mnl50.htm)
4. Oil recovery utilizing acidic organic phosphate scale inhibitors (1972), United States Patent 3703928 Website: http://www.freepatentsonline.com/3703928.html
5. Prevention of calcium sulfate scale in a waterflooding process (1971), United States Patent 3584686 Website: http://www.freepatentsonline.com/3584686.html
6. Gordon M. Graham, G.M., Eric James Mackay, E.J., and Sarah J. Dyer, S.J. (2002), 'The Challenges for Scale Control in Deepwater Production Systems: Chemical Inhibition and Placement', *NACE International*, Corrosion Conference, Paper No. 02316, April 7-12, Denver, CO, USA.
7. Salman, M., H. Qabazard, H., Moshfeghian, M.(2007),'Water scaling case studies in a Kuwaiti oil field', *Journal of Petroleum Science and Engineering*, Vol. 55,No. 1-2, January, pp. 48-55
8. Jamialahmadi, M. and Müller-Steinhagen, H. (2008) 'Mechanisms of scale deposition and scale removal in porous media', *Int. Journal of Oil, Gas and Coal Technology*, Vol. 1, Nos. 1/2, pp.81–108.
9. Fadairo, A.S.A. and Falode, O. (2009) 'Predicting tool for sulphate scale build-up around the wellbore in oilfield', *Int. Journal of Oil, Gas and Coal Technology*, Vol. 2, No. 4, pp.347 -364.
10. Dalvi, A.G.I., Kither Mohammad, N.M., Al-Sulami, S., Sahul, K. and Al-Rasheed, R., (2000) 'Role of chemical constituents in recycle brine on the performance of scale control additives in MSF plants', *Desalination*, Vol. 129, No. 2, pp.173-186.
11. Akbulut, U. and Hacioğlu, B. (1991) 'Electronitiated polymerization of maleic anhydride by direct electron transfer', *Journal of Polymer Science*, Part A: Polymer Chemistry, Vol. 29, No. 2, pp. 219-224.
12. '*Scale Inhibitors*' (1997) , Introduction of ACUMER series polymers, Rohm and Haas Company, http://www.rohmhaas.com , March 1997 Report, Taken from website: http://www.e-chemicalsource.com/bulletin/Acumer1000_1020.pdf
13. Shakkthivel, P., Sathiyamoorthi, R. and Vasudevan, T. (2004) 'Development of acrylonitrile copolymers for scale control in cooling water systems', *Desalination*, Vol. 164, pp.111-123.
14. Shakkthivel, P. and Vasudevan, T. (2006) 'Acrylic acid-diphenylamine sulphonic acid copolymer threshold inhibitor for sulphate and carbonate scales in cooling water systems', *Desalination*, Vol. 197, pp.179-189.
15. Li, H.-Y., Ma, W., Wang, L., Liu, R., Wei, L.-S., and Wang, Q. (2006) 'Inhibition of calcium and magnesium-containing scale by a new antiscalant polymer in laboratory tests and a field trial', *Desalination*, Vol. 196, No. 1-3, pp. 237-247.
16. Wang, C., Li, S.-P., and Li, T.-D. (2009) 'Calcium carbonate inhibition by a phosphonate-terminated poly(maleic-co-sulfonate) polymeric inhibitor', *Desalination*, Vol. 249, pp.1-4.
17. Tang, W., Yang, W., Yin, X., Ying Liu, Y., Yin, P. (2008) 'Investigation of CaCO3 scale inhibition by PAA, ATMP and PAPEMP', *Desalination*, Vol. 228, pp.55-60.
18. Al-Roomi, Y. M., and Hussain, K.F. (2005) 'Free Radical Solution Copolymerization of n-Octylacrylamide and Maleic Anhydride in a Non Polar Solvent; Synthesis and Characterization of a Difunctional Copolymer', *Polymer Bulletin*, Vol. 55, pp. 81-93.

19. Al-Roomi, Y. M., and Hussain, K.F. (2006) 'Homo-oligomerization of maleic anhydride in nonpolar solvents: A kinetic study of deviations from nonlinear behavior', Journal of *Applied Polymer Science*, Vol. 102, No. 4, pp. 3404 -3412.
20. ASTM International, West Conshohocken, PA, USA. http://www.astm.org/
21. NACE International (The *National Association of Corrosion Engineers*) Standards, Houston, TX, USA. http://www.nace.org/

(524b) A Novel Design for in-Flow Deposition Property Measurement

Mark Jemmett, Milind Deo, J. J. Magda, Husam El Gendy, Preston Montalvo and Dave Wagner, Chemical Engineering, University of Utah, Salt Lake City, UT

While wax deposition is a fairly well-known phenomena due to chemical potential gradients and solid-liquid equilibrium, effectively visualizing and measuring the thickness and strength of the deposition formed during flow without disturbing the crystalline structure is very difficult. Members of the department of Chemical Engineering at the University of Utah have constructed a flow loop that allows for non-invasive exploration of many facets of wax deposition. This loop includes an all-steel experimental heat exchanger where controlled deposition takes place under specified heat fluxes including "cold flow" – slurry conditions (below WPT), and zero heat-flux through the walls (isothermal) – and a clear acrylic plexiglass exchanger for controlled deposition and PIV visualization. Both sections are interchangeable, use pure ethylene glycol coolant, and also use high-resolution pressure transducers to measure pressure drop across the exchanger, allowing for accurate thickness measurements to be calculated using a modified Hagen-Poiseuille equation. In addition, the novel flow loop design incorporates a large, temperature-controlled reservoir to maintain fluid properties, as well as a particle size analyzer for use in determining particle size effect on deposition rate, strength, etc, and a conditioning loop to control these sizes. A method for determining in-line yield stress is also being developed so as to not break the crystals by transport out of the temperature controlled loop. The method currently considered is a pneumatic piston that fits flush inside the test section and will apply shearing stress until the deposition yields. Rheology of oils is extensively explored. Overall the flow loop design is unique, robust, and provides some of the most extensive information regarding the study of deposition available. Data generated shows measurement techniques within high accuracy of visual measurements. Novel design is also useful for any oil type, and is designed to handle water. Water effects to be explored after complete diagnostics and plain-oil exploration complete.

See more of this Session: Chemistry, Thermodynamics and Multiphase Flow of Complex Fluids Related to Flow Assurance II
See more of this Group/Topical: Topical D: Chemical Engineering in Oil and Gas Production and Other Complex Subsurface Processes

(524c) The Effect of Operating Temperatures and Solubility Curve of Paraffin On Wax Deposition

Zhenyu Huang[1], Shanpeng Han[1], Rainer Hoffmann[2] and H. Scott Fogler[1], (1)Chemical Engineering, University of Michigan, Ann Arbor, MI, (2)Herøya Research Centre Statoil ASA, Porsgrunn, Norway

Wax deposition in the subsea pipelines has been a major flow assurance problem for the petroleum industry. Flow loop experiments with different operating temperatures can be used to simulate the wax deposition at different axial positions of subsea pipelines. In this work, a series of flow loop deposition experiments are carried out based on different inlet temperatures and a model is applied based on the fundamentals of heat and mass transfer. The experimental results show that the wax deposit thickness increases with decreasing inlet temperature, which presents a contradictory trend from a previous study based on model oil [1]. To investigate this discrepancy, theoretical analysis using fundamentals of transport phenomena in the deposition model was carried out on the experiments of both the model oil and the crude oil. According to the deposition model, the temperature gradient at interface and the change in the solubility of wax at interface temperature are the two major elements to affect the growth of the wax deposit. By analyzing these two elements in the two aforementioned experiments, it is found that the discrepancy is due to the difference in the crude oil property and operating conditions. In conclusion, no generalization can be made to estimate on the effect of inlet temperature on wax deposition without considering the change in the temperature gradient of the particular experiment and the slope of the solubility curve of the oil.

[1] Hamid, O. B., Mehrotra, A. K. (2009) Energy & Fuels 23 3184-3194

See more of this Session: Chemistry, Thermodynamics and Multiphase Flow of Complex Fluids Related to Flow Assurance II
See more of this Group/Topical: Topical D: Chemical Engineering in Oil and Gas Production and Other Complex Subsurface Processes

(524d) The Propagation of Pressure and Wall Shear Stress in a Gelled Laboratory Pipeline with Application to the Flow Assurance of Paraffinic Crude Oils

Husam M. El Gendy[1], Mataz Alcoutlabi[1], Mark Jemmett[1], Milind Deo[1], Jules Magda[1], Rama Venkatesan[2] and Alberto Montesi[2], (1)Chemical Engineering, University of Utah, Salt Lake city, UT, (2) Chevron Technology Company, Houston, TX

There are many circumstances that lead to shutting down of oil pipelines for operational or emergency reasons in locations with ambient temperatures below the wax appearance temperature.Under these static conditions, paraffinic waxes in crude oils start to precipitate and form gels. Restarting the gelled pipeline, and eventually clearing the gel, requires the application of a pressure upstream of the gel that gives rise to a shear stress that exceeds the gel yield stress value.Pressure profile measurements and particle imaging velocimetry (PIV) experiments were performed both during gelation and after a sudden application of the restart pressure on a transparent pipeline containing a model waxy oil with measured rheological properties.

See more of this Session: Chemistry, Thermodynamics and Multiphase Flow of Complex Fluids Related to Flow Assurance II
See more of this Group/Topical: Topical D: Chemical Engineering in Oil and Gas Production and Other Complex Subsurface Processes

AUTHOR INDEX

AUTHOR INDEX